AMERICA
the Vulnerable

AMERICA the Vulnerable

How Our Government Is Failing to Protect Us from Terrorism

Stephen Flynn

HarperCollins*Publishers*

In cooperation with the
Council on Foreign Relations

HarperCollins books may be purchased for educational, business, or sales promotional use. For information, please write: Special Markets Department, HarperCollins Publishers Inc., 10 East 53rd Street, New York, NY 10022.

A portion of this work previously appeared in *Foreign Affairs*.

Founded in 1921, the Council on Foreign Relations is an independent, national membership organization and a nonpartisan center for scholars dedicated to producing and disseminating ideas so that individual and corporate members, as well as policymakers, journalists, students, and interested citizens in the United States and other countries, can better understand the world and the foreign policy choices facing the United States and other governments. The Council does this by convening meetings; conducting a wide-range Studies program; publishing *Foreign Affairs*, the preeminent journal covering international affairs and U.S. foreign policy; maintaining a diverse membership; sponsoring Independent Task Forces; and providing up-to-date information about the world and U.S. foreign policy on the Council's website, www.cfr.org.

FIRST EDITION

Designed by Nancy Singer Olaguera

Printed on acid-free paper

Library of Congress Cataloging-in-Publication Data

Flynn, Stephen.
America the vulnerable : how our government is failing to protect us from terrorism / by Stephen Flynn.—1st ed.
p. cm.
ISBN 0-06-057128-4
1. Terrorism—United States. 2. Terrorism—United States—Prevention. I. Title

HV6432.F58 2004
363.32'0973—dc22 2004047426

04 05 06 07 08 v/RRD 10 9 8 7 6 5 4 3 2 1

To JoAnn and Christina, with love

Contents

Preface

America remains dangerously unprepared to prevent and respond to a catastrophic terrorist attack on U.S. soil. That was the chilling conclusion of an independent, bipartisan task force formed by the Council on Foreign Relations, which included two former secretaries of state, two former chairmen of the Joint Chiefs of Staff, a former director of the CIA and FBI, three Nobel laureates, and co-chaired by former Senators Gary Hart and Warren B. Rudman, who previously led the U.S. Commission on National Security. I served as the director and lead author of the report, which was released in October 2002.

This book goes to press eighteen months after this blue-ribbon panel released their sobering assessment under the title *America—Still in Danger, Still Unprepared.* Despite the passage of time, our state of homeland insecurity has not materially changed.

U.S. soldiers continue to make the ultimate sacrifice in our war on terrorism overseas. Meanwhile, Americans have been reluctant to take the pragmatic measures, which our affluence can well afford, to address vulnerabilities at home.

For two centuries, geography has been America's biggest

security asset. With oceans to the east and west and friendly neighbors to the north and south, the United States has been untrammeled by enemy boots on our ground. Inhabiting the most peaceful corner of the world has meant that captains of industry and urban planners have been able to treat security as a marginal issue. Those carefree days are now gone and unfortunately we have inherited critical infrastructures so open that they offer terrorists a vast menu of soft targets.

The president, Congress, governors, and America's city mayors have taken some helpful measures to address some of our most glaring problems, particularly in how the government organizes itself to tackle the monumental task of improving our security. Unfortunately, we will not see the full effect of these post–9/11 initiatives for some time to come. While it is unrealistic to expect that we can eliminate, overnight, vulnerabilities that have been decades in the making, we must do better.

Throughout our history, Americans have displayed an extraordinary degree of resolve, nimbleness, and self-sacrifice in times of war. Today we are breaking with that tradition. Our nation faces grave peril, but we seem unwilling to mobilize at home to confront the threat before us. Managing the danger that al Qaeda poses cannot be achieved by relying primarily on military campaigns overseas. There are no fronts in the war on terrorism. The 9/11 attacks highlighted the fact that our borders offer no effective barrier to terrorists intent on bringing their war to our soil. Nor do their weapons have to be imported, since they have proven how easy it is to exploit the modern systems we rely upon in our daily lives and use them against us.

My sense of foreboding about America's current state of vulnerability predates 9/11. It has been welling up throughout my twenty-year career as a commissioned officer in the U.S. Coast Guard, which I spent as a sailor, college professor, and public-policy practitioner. In two commands at sea, I experienced first-hand how difficult it is to police our borders. While a teacher at the U.S. Coast Guard Academy and on fellowships to the Brookings Institution, the Annenberg School for Communication at the University of Pennsylvania, and the Council on Foreign Relations, I became preoccupied with the growing capabilities of criminals and terrorist organizations. I discovered how intransigent the federal bureaucracy could be to adapting to new threats while I was a director on the National Security Council staff and as an advisor for the U.S. Commission on National Security. When the attacks of September 11 came, I shared the feelings of horror that all Americans felt. But like most of the small cadre of individuals who had been keeping their eyes on al Qaeda, I was not surprised.

In 2000, I wrote an essay published by *Foreign Affairs* which included a scenario describing how Osama bin Laden might exploit our perilously exposed transportation system to smuggle and detonate a weapon of mass destruction on our soil. The article led to invitations to conduct briefings around Washington, D.C. In those pre–9/11 presentations I maintained that these attacks would involve more than the loss of innocent lives, but would generate a spasmodic response to shut down the entire transportation sector as public officials struggled to determine what happened. This would be followed by a rash of poorly conceived new security mandates in the scramble afterward to reassure an anxious American public. The resultant economic and societal disruption would be precisely the kind of outcome that groups like al Qaeda aspire to achieve. My conclusion was straightforward. Given the enormous stakes, we should make transportation security a priority before terrorists strike.

These briefings proved to be enormously frustrating. Even those people who understood al Qaeda's commitment to attacking America were generally reluctant to recognize the degree to which our guard was down at home. In the absence of specific intelligence, most policymakers were unwilling to acknowledge the threat to our transportation infrastructure was real. At the same time, the intelligence community was dedicating virtually no resources to assessing a threat posed by terrorists and criminals to ships, trains, trucks, planes, and containers. Those few who shared my concern were convinced that little could be done. The common refrain I heard was, "Americans need a crisis to act. Nothing will change until we have a serious act of terrorism on U.S. soil."

Sadly, it turns out that even 9/11 has not served as a catalyst for the United States to take stock of its many vulnerabilities. Over the past two and a half years, I have traveled extensively around the nation, meeting with frontline agents, first responders, law enforcement officials, and civic groups. What I have found are pockets of innovation by dedicated public servants who are being tasked each day to do the impossible: to secure a nation that has not been mobilized to defend itself at home.

The United States has failed to secure its homeland. Many in Washington will likely protest this judgment, arguing that I do not adequately account for the progress they are making. But earning a passing grade is not only about making an effort. Just as any student knows he will fail an exam if he answers forty questions correctly yet leaves sixty blank, the measures that have been taken to protect the homeland must be judged against a standard that assesses their adequacy to handle the threat and the consequence, should those protective measures fail. I know how exposed our nation was prior to 9/11, and I have been closely monitoring what has been happening to improve our security since that tragic day. There are too many incompletes to justify a passing grade.

Critiquing where we are is a necessary stepping-off point for outlining where we must go. Every day that there is not an attack, Americans become increasingly wistful for a sense of pre–9/11 normalcy. Perky reassurances by public officials that they have matters well in hand are not only inaccurate, but they remove the oxygen from a sustained effort to confront the ongoing terrorist threat. The war on terrorism is not one that can be permanently won. The means to conduct terrorist acts are too cheap, too available, and too tempting ever to be completely eradicated. Terrorism is a threat that we must constantly combat if we are to reduce it to manageable levels so that we can live normal lives free of fear. September 11 marked the end of an era during which we could go about our lives treating security as something only other people had to worry about.

But security need not become our sole preoccupation. In fact, many of the most effective tools for combating the terrorist threat often can provide other important societal benefits. Invigorating our public health services to identify and respond to a bio-threat will help them to manage naturally occurring diseases like SARS and Avian influenza. Equipping our emergency responders to communicate in the wake of a terrorist attack will make them better prepared to save lives in any natural or man-made disaster. Bolstering frontline agents to detect and intercept terrorists will strengthen their hand in combating criminals.

There are countless examples where pragmatic steps taken to reduce our exposure to terrorism can be sound investments, even in the absence of the terrorist threat that now confronts us.

This book proposes a new framework for how we should deal with the post–9/11 world. It offers pragmatic suggestions on how to tackle specific vulnerabilities, including emergency responder issues, and in the areas of trade, transportation, and border security. This topic could certainly benefit from drawing on a wider pool of scholarship than now exists and on the kind of perspective that only the passage of time could provide. But time is not on our side.

Good detailed scholarship will not emerge in a vacuum. A demand for it must come first. If I have erred on the side of being succinct and provocative, it has not been out of the confidence that I know all the answers, but out of a sense that the issues deserve far more attention and serious thinking than they are receiving. I welcome the marshalling of new data to show that the conclusions I have drawn are under- or overstated. There is little risk for the nation if it turns out that the world is more benign and our government is better prepared than I believe it to be. The real danger lies in placing too much confidence in security measures that our tenacious enemies can readily evade.

In the fall of 1999, I was invited to join the studies program at the Council on Foreign Relations and arrived in New York with the goal of researching and writing a book on how border management must adapt to the imperative to more effectively police people and goods moving at greater speeds through the international system. I was well on my way to completing that task when the 9/11 attacks took place, making that project much more relevant, but also unleashing changes that made much of my data immediately out of date.

Writing was again put on hold as I soon found myself testifying before Congress, conducting briefings around Washington, participating in a team that drafted the Coast Guard's Maritime Homeland Security Strategy, advising the Commissioner of Customs on what became the Container Security Initiative, serving on a National Academy of Sciences Advisory Panel on Transportation Security, and jump-starting two private-public container security programs: Operation Safe Commerce, and Smart and Secure Trade Lanes. Then I was asked to serve as project director for the Homeland Security Task Force in the summer and fall of 2002. Once that was completed, I received a book contract from HarperCollins which came with the windfall of having Tim Duggan as my editor. Tim has been a stellar coach, working hard to wean me off academic and policy-wonk prose and to sharpen the focus and message of this work.

In addition to my editor, there are many people who have helped to make this book possible. At the top of that list are the three research assistants who have worked with me since 9/11. Sean Burke was with me in my New York office when the first plane hit, was at my side when I visited with the emergency responders at Ground Zero, and has since stood by me through thick and thin. I am so very grateful that he has been willing to channel his abundant talents to supporting my work these past four years. Robert Knake is another gifted young man and an extremely talented researcher. Rob's keen mind is matched only by his wonderful wit, which has been a welcome relief, given the gravity of the material. Daniel Dolgin has been indispensable to me throughout the final stretch, lending his considerable energy and intellect to the project. Dan, Rob, and Sean are easily among the best and brightest of their generation and it has been an honor to be in their good company.

I have had some wonderful colleagues during my Coast Guard career who consistently shattered the stereotype that the military does not provide aid and comfort to thinkers. My first fellowship came about thanks to David Long, Joseph Egan, Irving King, Earl Potter, and William Sanders who conspired to allow me to become the first Coast Guard officer serving on the faculty at the Coast Guard Academy to be selected as a Council on Foreign Relations International Affairs fellow. My tenure at the Council during my last years as an active duty Coast Guard officer was made possible due to the support of Robert Ayer and Nils Wessell; Coast Guard Academy Superintendents Douglas Teeson and Robert Olson; and, at Coast Guard Headquarters, Admirals James Loy, Timothy Josiah, Patrick Stillman, and David Nicholson.

I am particularly indebted to Leslie Gelb, now president emeritus at the Council on Foreign Relations. It was Les who arranged in 2000 for me to work with the support staff for the Hart-Rudman Commission on National Security. He also raised the funds to establish a chair named in Ambassador Jeane J. Kirkpatrick's honor, and recommended to the board that I become its inaugural occupant upon my retiring from the Coast Guard in 2002. The highlight of my time at the Council has been the chance to direct the task force with Senator Rudman and Senator Hart, which was a project that Les made both his personal priority and a top priority for the Council. He also introduced me to America's top literary agent, Morton Janklow, who graciously agreed to take me on as a client. Finally, Les has lent his time, intellect, and sharp editorial eye to this project from start to finish.

Peter Peterson and Carla Hills, the chairman and vice chairman of the board at the Council on Foreign Relations, have been very generous supporters of my work, offering me their sage counsel at critical points. Also at the Council, I was fortunate to work for Lawrence Korb, who served as director of studies for much of the time I was working on this project. Larry's door was always open for me, and I wore a path to it on good days and bad. I always left his office with a skip in my step. James Lindsay came aboard to replace him in the fall of 2003 and generously reviewed my manuscript and offered me many useful comments. Richard Haass has taken over the leadership reins at the Council and has given me the benefit of his powerful intellect and sharp editorial eye. David Long, Jordon Pecile, Maryann Cusimano-Love, Sidney Wallace, Rey Koslowski, Lawrence Wein, John Holmes, Peter Boynton, Robert Kelly, Larry Kiern, Robert Castle, George Haynal, and Peter Hall also looked over the manuscript and provided me with excellent suggestions to improve it. The remaining shortcomings are strictly due to my own limitations.

My engagement on this homeland security issue owes its origins to General Charles Krulak, whom I first worked for in the George H. W. Bush White House, and Richard A. Clarke, who made me a part of his office staff on the National Security Council in the Clinton White House. While there was no shortage of revelers at the end-of-history party in the 1990s, Chuck Krulak and Dick Clarke kept a vigilant and wary eye out for the gathering storm. They brought all their formidable leadership and bureaucratic skills to bear to get the national security community to prepare for its arrival. The embryonic preparations America had made to deal with this terrorist threat prior to 9/11 were in no small part due to their Herculean efforts.

Senator Rudman and Senator Hart lent their considerable intellects and statesmanship to the pre–9/11 cause of waking up America to the danger of catastrophic terrorism. They were supported by General Charles Boyd, who served as executive director of the U.S. Commission on National Security. It was a distinct honor to work with my colleague Frank Hoffman to support these outstanding Americans as they wrestled with the issue of how to organize our national-security establishment to protect the homeland.

I am haunted by America's persistent state of vulnerability. I live each day cursed with an awareness of how far we have to go to protect our nation. My dear wife, JoAnn, and daughter, Christina, have had to share the burden of this preoccupation. They have put up with a husband and father who has been on the road too often, or holed up in the study writing. I am so very grateful for their bountiful love. Without their unwavering support, I would have faltered in this effort long ago.

AMERICA
the Vulnerable

1

Living on Borrowed Time

If September 11, 2001, was a wake-up call, clearly America has fallen back asleep. Our return to complacency could not be more foolhardy. The 9/11 attacks were not an aberration. The same forces that helped to produce the horror that befell the nation on that day continue to gather strength. Yet we appear to be unwilling to do what must be done to make our society less of a target. Instead, we are sailing into a national security version of the Perfect Storm.

Homeland security has entered our post–9/11 lexicon, but homeland insecurity remains the abiding reality. With the exception of airports, much of what is critical to our way of life remains unprotected. Despite all the rhetoric, after the initial flurry of activity to harden cockpit doors and confiscate nail clippers, there has been little appetite in Washington to move beyond government reorganization and color-coded alerts.

While we receive a steady diet of somber warnings about potential terrorist attacks, the new federal outlays for homeland security in the two years after 9/11 command an investment equal to only 4 percent of the Pentagon's annual budget. Outside of Washington, pink slips for police officers and firefighters are more common than new public investments in security. With state and local budgets hemorrhaging red ink, mayors, county commissioners, and governors are simply in no position to fill the security void the federal government has been keen to thrust upon them. The private sector has shown its preference for taking a minimalist approach to new security responsibilities. There have been private-sector leaders who have been bucking this trend, several of whom are featured in the pages ahead. But, by and large, trade and industry associations have been hard at work trying to fend off new security requirements that might compel them to address vulnerabilities and thereby raise their bottom-line costs.

From water and food supplies; refineries, energy grids, and pipelines; bridges, tunnels, trains, trucks, and cargo containers; to the cyber backbone that underpins the information age in which we live, the measures we have been cobbling together are hardly fit to deter amateur thieves, vandals, and hackers, never mind determined terrorists. Worse still, small improvements are often oversold as giant steps forward, lowering the guard of average citizens as they carry on their daily routine with an unwarranted sense of confidence.

Old habits die hard. The truth is America has been on a hundred-year joyride. Throughout the twentieth century we were able to treat national security as essentially an out-of-body experience. When confronted by threats, we dealt with them on the turf of our allies or our adversaries. Aside from the occasional disaster and heinous crime, civilian life at home has been virtually terror-free. Then, out of the blue, the 9/11 attacks turned our national security world on its head. Al Qaeda exposed our Achilles' heel. Paradoxically, the United States has no rival when it comes to projecting its military, economic, and cultural power around the world. But we are practically defenseless at home.

A number of post–Cold War realities have created a new global environment that places America in a position of especially grave danger. First, from nearly all points on the compass, there is rising anti-Americanism. To a large extent this is the inevitable byproduct of the United States' unique standing as the sole remaining superpower. Our current predicament is that any unhappy person on the planet is inclined to lay the blame on America's doorstep. If they think their society is being undermined by cultural pollution, they are likely to see the United States as the lead polluter. If they view the economic rules of the game as rigged to benefit the few at the cost of the many, they castigate the United States as the capitalist kingmaker. And if they imagine life would be better if there were a change in the local political landscape, they see the United States as standing in the background, or foreground, as a barrier to their revisionist dreams. When our actions and policies display periodic arrogance and indifference, we only add grist to the anti-U.S. mill.

For the foreseeable future, increased anti-Americanism will be a fact of life. Certainly it can be exacerbated or ameliorated by the approaches we take and the priority we assign to addressing some of the world's most pressing public-policy challenges. There is too much pent-up rage and frustration around the world caused by overpopulation, limited education and job opportunities, and a lack of participation in the political process. As a nation, we should be mindful of these sobering realities and work to improve them wherever we can. But even the wisest, kindest, and gentlest American leadership will not appease groups like the remnants of the Taliban, whose beliefs are the antithesis of our own. There will be ample recruits to strike out at the United States as a means of defending or advancing their causes.

This rise in discontent is made more menacing by another disturbing fact of twenty-first-century life: groups with no governmental ties can acquire the most lethal tools of warfare. Certainly, state sponsorship can be helpful. But with so many pockets of the world hosting open-air arms bazaars, complicity with an established government is not essential. At one end of the spectrum, weapons like the AK-47 are so plentiful that they can be had for the price of a chicken in Uganda, the price of a goat in Kenya, and the price of a bag of maize in Mozambique or Angola. At the other end, there is enough separated plutonium and highly enriched uranium in the world to make thousands of nuclear weapons. Weapons-usable nuclear materials exist in over 130 research laboratories operating in more than forty countries around the world, ranging from Ukraine to Ghana.

Added to the growing motive and means to strike out at the United States, there is also enhanced opportunity in our interconnected global environment for criminals and terrorists to acquire wider reach. Over the past decade, drug smugglers, human traffickers, and gunrunners have found fewer barriers to their nefarious activities. As the worldwide networks that support international trade and travel become more open and the level of cross-border activities increase, the bad have benefited alongside the good.

And it is not just terrorists who would enjoy fewer barriers to successfully targeting the United States. Like moths to the flame, our current and future enemies will find the opportunity irresistible to assault nonmilitary elements of U.S. power that arise from our growing dependence on sophisticated networks to move people, food, cargo, energy, money, and information at higher volumes and greater velocities. For years, this mounting dependency has not been matched by a parallel focus on security. The architects of these networks have made efficiency and diminishing costs their highest priority. Security considerations have been widely perceived as annoying speed bumps in achieving their goals. As a result, the systems that underpin our prosperity are soft targets for those bent on challenging U.S. power.

Ironically, our overwhelming military capabilities make it attractive to target the nonmilitary backbone of our power. We spend more on conventional military muscle than the next thirty countries combined. Never in human history has such a disproportionately large amount of force been concentrated in the hands of one nation. In the face of this reality, our adversaries must become creative Davids to our Goliath.

Our situation is not without precedent in world history. When General Charles C. Krulak served as Commandant of the U.S. Marine Corps from 1995 to 1999, he often recounted the fate of Roman Proconsul Quintilius Varus to suggest what may befall us if we fail to adapt to the kind of warfare our enemies will be waging in the twenty-first century.

Varus was serving as governor in Northern Europe in 9 A.D. when he received word that Roman outposts north of the Rhine had come under attack by rebellious Germanic tribes. With around 15,000 experienced and well-armed Roman soldiers at his disposal, putting down the barbarians should have been a straightforward task. Just three years earlier, the Romans had easily squelched a similar uprising that resulted in the death or enslavement of thousands of Germanic tribesmen and their women and children. But this time the tribes didn't follow the old script. They knew they could not match the power of the Roman heavy cavalry and archers in the open field. So when Varus led his legions into Teutoburger Forest on a hot August morning, the tribesmen set upon them in the trees, banking on the fact that close combat would neutralize the lethal Roman cavalry and bowmen. It worked. Within three days, Varus had lost the eagles of three legions, and shortly thereafter, his head adorned a Germanic war pike. Fewer than one hundred Romans were reported to have survived the battle.

Krulak would tell his audiences that we should take two lessons to heart from the Roman debacle: "First, a symmetrical and overwhelming approach by a dominant power always invites an asymmetrical counter. Second, Roman strategy proved to be inflexible and resistant to change. As a superpower, Rome never came to grips with the 'shadow war' waged by the Germanic tribes."

Today, harnessing terror to target our civilian foundation promises excellent odds of inflicting serious damage to the United States. Our economic engine makes it possible to not only bankroll the hard power of our second-to-none military forces, but it allows us to project the soft power that flows from being home to the planet's premier marketplace and dominant culture. Since the barriers to sabotaging that engine are few, by attacking the soft underbelly of globalization, our enemies erode our ability to actively pursue our interests around the world. They also end up undermining the global economy at large.

The chink in our superpower armor has become apparent to many. Five years before 9/11, Osama bin Laden highlighted the nexus between American military power and its economy in his first fatwa, "The Declaration of War Against the Americans Occupying the Land of the Two Holy Places." In that manifesto, bin Laden, who holds a degree in economics from the King Abdul-Aziz University in Jeddah, Saudi Arabia, called for a boycott of American goods as a way of "hitting and weakening the enemy." Two years later, in a joint statement with four other terrorist leaders announcing "Jihad Against Jews and Crusaders," he called on every Muslim to kill Americans and "plunder their money wherever and whenever they find it." While it is unlikely that he expected the events of 9/11 to wreak the economic damage that they did, there is little question that he both understood and drew solace from the aftershocks of the attacks. On October 21, 2001, in a broadcast that aired on Al-Jazeera, bin Laden remarks on how his attacks generated billions of dollars in losses to Wall Street, in the daily income of Americans, in building costs, and to the airline industry:

> I say the events that happened on Tuesday 11 September in New York and Washington, that is truly a great event in all measures, and its claims until this moment are not over and are still continuing. According to their own admissions, the share of the losses on the Wall Street market reached sixteen percent. . . . They have lost this, due to an attack that happened with the success of Allah

> lasting one hour only. The daily income of the American nation is $20 billion. The first week they didn't work at all due to the psychological shock of the attack, and even until today some don't work due to the attack. . . . The cost of the building losses and construction losses? Let us say more than $30 billion. Then they have fired or liquidated until today from the airline companies more than 170,000 employees.

One should assume that others also recognize that bin Laden's investment in terror paid remarkable dividends.

The case for broadening warfare to include nonmilitary targets in the United States was cogently made by two senior Chinese military officers at the century's end. Colonels Qiao Liang and Wang Xiangsui penned a book published in February 1999 on military strategy, which they boldly titled *Unrestricted Warfare*. In their treatise, Qiao and Wang propose tactics to compensate for the military superiority of the United States. They argue that the new principles of war have moved from "using armed force to compel the enemy to submit to one's will," and toward "using all means, including armed force or nonarmed force, military and nonmilitary, lethal and nonlethal to compel the enemy to accept one's interest." Among the tactics they advocate are hacking into Web sites, targeting financial institutions, and attacking the electrical grid and telecommunications networks.

The reason that catastrophic terrorism holds out such potential as a means to wage war on the United States is not simply because these attacks can inflict damage to systems we depend on; it is because our enemies have good reason to believe that a successful act of terror on American soil will trigger a reaction in which the U.S. government exacerbates localized destruction with substantial self-inflicted national and even global costs. Recall that in the case of 9/11, nineteen men with box cutters were able to accomplish what no other world power could dream of achieving by conventional means. Within hours of the attacks, our government severed our transportation umbilical cord with the world. As a result, we effectively imposed a blockade on our own economy.

The federal authorities reacted this way for two reasons. First, given the uncertainty surrounding the attacks, throwing a transportation kill switch was the only way they believed they could prevent other concurrent attacks that might be underway. Second, in order to restore public confidence that they could detect and intercept a follow-up attack, authorities had to demonstrate that frontline agencies had the means to police incoming people and goods. The only option for accomplishing that Herculean task was to slow legitimate traffic to the point where their agents might be able to actually examine it—and the consequence was gridlock. Elegantly and simply, the terrorists proved themselves to be exceptional students of American culture.

It is the challenges associated with managing uncertainty and restoring public confidence that practically guarantee that acts of terror will be a potent weapon of mass disruption. Once an incident occurs, government officials will almost certainly be on the defensive. This is because of the unique effect terror has on the American public's psyche. People reflexively put themselves in the shoes of the victims. If it could have happened to them, why can't it happen to us? While people know that their government can't prevent natural disasters, they do expect their officials to be vigilant in preventing our enemies from killing innocent civilians, toppling our landmarks, and destroying non-military property. Accordingly, there is a political price to be paid if politicians are perceived as being negligent or ineffectual in providing security. So when an act of terror takes place, it triggers a powerful political dynamic to leap to decisive protective actions before evaluating the likely costs or consequences. The greater the exploited vulnerability, the more likely the government is to overreach in response. Our enemies will exploit this dynamic.

Knowing that we face a future where our adversaries will have the motives, the weapons, and the opportunities to strike at and profoundly disrupt our nation, we should be preparing for the worst. But we are not. Moving from where we are to where we have to be will require a far more spirited and informed national conversation about the ends and means of homeland security than is currently underway. To date, that conversation has been constrained by the fact that the most vocal protagonists have put forth three simplistic positions.

At one extreme are the security-at-any-cost advocates who hold that we should pay any price to eliminate the possibility that terrorists can again strike us on U.S. soil. This zero-tolerance perspective makes security an end in itself and is often dismissive of any potential trade-off issues.

At the other extreme is the cure-is-worse-than-the-disease school which maintains that "the terrorists have won" if we pursue any new security measures that potentially constrain our pre–9/11 freedoms or raise the cost of doing business. From this libertarian perspective, Americans should oppose on principle all security initiatives that might intrude on the lives of individuals or the marketplace.

Then there is the escapist view that sidesteps homeland security altogether by advocating a go-to-the-source approach. This is the prevailing view held by the White House, at the Pentagon, and among many around Washington. It is easy to understand its popularity. On its face, the logic of stopping terrorists abroad before they can strike us at home is hard to contest. It also has the added benefit of allowing us to stick with our traditional approach to national security where threats are managed far from our shores so we can go about our daily business unencumbered by security imperatives at home. These advocates profess that costly investment in defenses can be avoided by bankrolling stepped-up intelligence collection and a nimbler military that can carry out preemptive strikes against those who would threaten us.

The one big problem with the "best defense is a good offense" mantra is that it manages to overlook the central reality of the 9/11 attacks. Al Qaeda has demonstrated its ability to establish a footprint on U.S. soil and successfully exploit the gaps associated with our long-standing division of labor between vigorously securing the nation abroad and tepidly securing it at home. And even overseas, there are limits to our power. It would be hubris to imagine that we can preempt all the schemes of violent young men who see America as their enemy and who live in the ghettos of cities like Karachi or Cairo. Nor are we in a position to police the remote villages in the Nile Valley or in Central Asia. It would seem that we are barely capable of hunting down these violent young men even when they are in our midst. And periodic displays of our overwhelming military power will not deter terrorists; in many cases, it will only bolster the ranks of new recruits.

We also need to keep in mind that for modern-day terrorists, our borders offer no real barrier to their slipping into the United States and carrying out a campaign of sabotage. This is because, for our frontline inspectors, border control remains the enforcement equivalent of trying to catch minnows at the base of Niagara Falls. In 2002, over 400 million people, 122 million cars, eleven million trucks, 2.4 million rail freight cars, approximately eight million maritime containers, and 59,995 vessels entered the United States at more than 3,700 terminals and 301 ports of entry. Given these numbers, obviously not everyone and everything gets checked when entering this country legally. U.S. customs and border-protection inspectors are charged with monitoring compliance with more than four hundred laws and thirty-four international treaties, statutes, and agreements on behalf of forty federal agencies. In general, lacking advance data, frontline agents have only thirty seconds for people and one minute to make a go/no-go decision for vehicles. And then there are the 7,000 miles of land borders and 95,000 miles of shoreline, which provide ample opportunities to walk, swim, or sail into the nation. Official estimates place the number of illegal migrants living in America at over seven million.

It is a sense of futility, fueled by the lack of vision about what sensible measures are worth pursuing, that lies at the heart of our national inertia on the homeland security issue. This book aims to provide an antidote for such an unhappy and dangerous state of affairs. Immediately following the attacks on September 11, I watched with frustration as the pendulum swung from what was essentially a token approach to security in airports to one where we rushed to put in place a system based on the notion that every passenger poses an equal risk of being a terrorist. There is an old security axiom: "If you have to look at everything, you will see nothing." Stated differently, dumb security can be as useless as no security. In fact, it can be more dangerous, since it creates the illusion of safety without actually providing it.

Identifying a way to optimize security while maximizing openness has been my preoccupation for the better part of ten years. A core obligation of government is to provide for the safety and security of its people. My concern has been that the imperative of openness has been trumping the public sector's means to meet that responsibility. Not only are innocent lives being placed in jeopardy, but the globalization process itself is at risk. In the face of a catastrophic terrorist threat, the general public will insist on protective measures. Following particularly traumatic incidents, the appeal of a neo-isolationist agenda may gather real traction. But it would be self-defeating for the United States to retreat from the world and hide behind a newly constructed twenty-first-century version of moats and castles. If we derail the engine of economic growth and retreat from Wilsonian values, which are indispensable to sustaining the promise and reality of a better life around the globe, we will end up fueling the threatening environment from which we are trying to protect ourselves. Security without openness is as self-defeating as openness without security.

The post–9/11 security imperative does not require making wholesale changes to our way of life that diminish our cherished freedoms. Nor does it boil down to bankrolling gold-plated measures to protect against every possible contingency. Instead, it is about addressing the economic costs and potential infringements on personal liberty in much the same way as we deal with those issues while managing more familiar dangers to modern life.

For instance, America has developed an enviable record on advancing industrial, automotive, and environmental safety while still maintaining the world's most powerful economy. At its heart, the safety agenda is about minimizing the chance that carelessness, accidents, or acts of God can destroy lives, damage property, and disrupt vital systems that we depend upon. This goal is widely supported by the general public even though it creates tensions. Safety requirements such as mandating the use of car seatbelts are constraints on individual choice. Requiring hotels to install fire sprinkler systems adds costs to the stays of business travelers and tourists. But we do not oppose these requirements as inimical to our way of life. Instead, we applaud efforts by legislators and regulators to develop meaningful incentives and sanctions to help strike the right balance between the pursuit of commercial and personal interests and the need to ensure that the public's safety is not placed at unnecessary risk.

In the same vein, our new security agenda should be about taking the appropriate steps to reduce the risk that our enemies can destroy and disrupt the things that we value. That translates into the need to develop the right carrots and sticks to encourage across-the-board security measures. This does not mean that the endgame is to turn every potential soft target into a fortress. More often, it involves improving the capacity to monitor systems and to develop contingencies for coping when something goes wrong. This is because terrorists want to be successful when they carry out an attack. They have limited resources and they are interested in achieving spectacular results. To this end, they will stake out their targets and if they discover the risk of detection is reasonably high, or that the damage from a successful attack can be quickly contained, they likely will go back to the drawing board.

This allows for something less than ironclad security. What is required is enough security to create a deterrent.

If we are smart in how we construct a security deterrent, we will achieve other benefits. For instance, the best way to protect our livestock industry from an act of bioterrorism is not to assign a cowboy with a shotgun to every cattle ranch. A better option is to develop an early-warning system to detect contaminated animals, and to have the means to quickly isolate all the livestock with which they may have come into contact. The good news is that putting in place this kind of capacity would accomplish more than fending off terrorists. As the discovery in December 2003 of a cow infected with mad cow disease highlighted, if such a system had been up and running, the U.S. government would have been in a better position to stave off the thirty-nation boycott of America's cattle export business, which earns over 3 billion dollars in annual sales. Encouraging the livestock industry to invest in measures that could lower exposure to the less likely scenario of a bioterror attack would help address the much higher-probability imperative that it will need to reassure consumers that the food supply is safe.

We need to take a deep breath and realize that, just as we have enhanced safety without turning ourselves into a police state, we can do the same with regard to security. Further, by making this investment up front, we can reduce the risk of expensive and dysfunctional overreactions when attacks do take place.

Stepping up to this vital agenda requires that Americans have a fuller appreciation of the danger in persisting with the status quo. Accordingly, I have made a point to write this book with candor. Admittedly, for someone who spent a career as a military officer, laying bare our ongoing state of national vulnerability runs counter to my guardian instincts. Also, it is hard to be the bearer of sober news on how little progress we are making when I know that tens of thousands of dedicated public servants have been working around the clock since 9/11, doing everything within their power to address the daily challenges that confront them.

Nonetheless, I learned from my eight years as a Coast Guard Academy professor, teaching the principles of American government to future officers, that a democracy will not make hard decisions unless its citizens understand both the facts and the stakes involved. The men and women on the front line will never receive the kind of resources they need to properly do their jobs if the public does not express a willingness to see tax dollars invested to that end.

While some readers may be shocked to learn just how insecure this country is, be assured that global criminals and terrorists are very much in the know. The vulnerabilities I point to are drawn from open-source information about actual criminal or terrorist acts. Highlighting how much work remains to be done is not the danger. The invitation to disaster is in persisting with a status quo in which public officials toss and turn at night, mindful of how exposed we remain, while the general population goes about their lives oblivious to the perils that confront them.

2

The Next Attack

The White House, the Pentagon, and the new Department of Homeland Security must assume that our enemies will soon launch far more deadly and disruptive attacks than what we experienced on September 11, 2001. The potential scenarios are almost unlimited. By now, we are all aware of the extent to which the materials for the construction of weapons of mass destruction are available, how criminal networks have been exploiting global transportation networks to move contraband, and how al Qaeda operatives have gained access to Western societies and successfully operate in our midst. One could easily conclude that a tenacious and unscrupulous enemy might assemble an attack that looks like the following:

One day, an al Qaeda operative may receive marching orders from his bosses to make his way to Ukraine to pick up a small shielded container of radioactive material. His mission will be to smuggle the material into Germany, where a bombmaker will add it to a conventional explosive device. Then the bomb will be loaded in a shipping container and sent to the United States.

Let's say his name is Omar and he grew up in the Punjab region of Pakistan, where he attended one of the infamous religious fundamentalist schools or madrassa, such as Jami` at Dar al `Ulum Haqqaniyah, located outside Peshawar. There he committed the Quran to memory and learned from his schoolmaster, Maulana ul-Haq, the radical jihadist version of Islam that holds that it is the moral obligation of a true Muslim to defeat evil and promote Islam by force against all non-Muslims.

Since jobs in the Pakistani countryside are scarce, once he completes his schooling he moves to Karachi to live with his uncle's family in the sprawling megacity of ten million people that serves as Pakistan's outlet to the sea. He finds work as a ship breaker, a dangerous third-world industry in which hulls of old merchant ships are driven up on a beach and dismantled by an army of men for the scrap metal. With the help of a cousin, he is able to escape this job and obtain the appropriate papers to become an able-bodied seaman on a tramp steamer. From the deck of that ship, Omar sees enough of the world to convince him that his teachers were right about the decadent West. When he returns to Karachi, he is an easy target for an old madrassa classmate—now an al Qaeda recruiter.

For the Ukrainian smuggling operation, Omar's merchant marine credentials are ideal. He will find a ship that makes port calls in Northern Europe, including his intended destination, Gdansk, Poland. Once Omar's ship is moored to its berth, the Polish immigration official will give only a cursory look at the crew-list visa and the sailor's identity papers before the seamen move down the gangplank. While Omar's shipmates make a beeline for the red-light district, he will catch a train to Warsaw.

There he uses an al Qaeda-provided Belgian passport—one of the nearly 20,000 blank documents that have disappeared from Belgian passport offices and consulates since 1990—to apply for a tourist visa at the Ukrainian consulate. At the border near the Polish town of Pzemysl, after a token inspection of his rental-car trunk, he is waved through by an uninspired and poorly paid Ukrainian border inspector in Doromyl.

Omar's next stop is the Black Sea port of Odessa. There he has a rendezvous in a nondescript warehouse where he receives the stolen radioactive material. The material is americium-241, which came from a stolen industrial device used in the Azerbaijani oil fields. Omar's job is to get the americium-241 to an al Qaeda terrorist cell operating out of Germany.

To accomplish this, he uses the vast river and canal system that serves as a waterborne highway throughout the heart of Europe from the Black Sea to the North Sea. Hundreds of river barges, known as "eurobarges," ply these waters, carrying a variety of bulk materials ranging from grain to heating oil. Most are literally family-run operations, with the women and children living aboard. The Danube River is the system's eastern branch beginning at the port of Constantza, Romania, 140 miles southwest of Odessa.

Smuggling has become a thriving industry in Odessa. In 2002, Ukrainian law enforcement authorities seized 1,200 metric tons of illegal narcotics and 2,000 metric tons of cocaine precursors in the port, but this was a drop in the bucket. Since the collapse of the Soviet Union, Ukraine has been on the western end of a well-established heroin transit route originating in Central and Southwest Asia. The combination of long, porous borders and woefully underfinanced and underequipped border-control agencies make getting drugs into Ukraine a nearly risk-free proposition. Moving them out of the country is even easier.

The bit of Russian that Omar picked up as a merchant seaman happens to serve him well during a meeting with a eurobarge captain who has just taken on a load of coal destined for Kelheim, Germany. To join the crew, Omar offers the captain $2,500 in advance and another $2,500 on arrival, letting on that he is a drug courier. For the river captain, dope is just another commodity that holds the added allure of being far more lucrative than what he earns from transporting traditional cargo.

Omar adapts easily to the vessel's routine, and his shipboard experience allows him to pass himself off as a deckhand. His package lies hidden away below decks underneath the spare mooring lines in an untidy storage compartment. The ten-day trip to Kelheim is uneventful. During the three occasions the eurobarge is hailed by customs officials along the way, the inspectors never get beyond the pilot house. The captain and vessel are familiar sights on the Danube. Besides, the work of opening rusty old hatches and clamoring around dark and dirty holds requires more physical labor than their paltry government salaries inspire.

The Bavarian city of Kelheim lies almost due north of Munich at the intersection of the Danube and the Gunzenhausen rivers. One of Omar's accomplices arrives from Frankfurt while the barge is unloading its coal. In the bustle of this backwater port, no one notices a deckhand carrying a duffle bag across the gangplank and loading it into the back of a van. The accomplice's specialty is bombmaking, having received a six-week course in Osama bin Laden's Afghan Darunta camp in 1999, the same camp where millennium bomber Ahmed Ressam trained in preparation for his planned attack on Los Angeles International Airport. He mixes the radioactive material into an array of conventional high-powered explosives. When the bomb is detonated, its explosive force will disperse americium in a cloud that will contaminate an area measuring over one mile long and covering the equivalent of sixty city blocks.

The two men's destination is Mannheim, which is home, among other large companies, to a 200-acre manufacturing plant for DaimlerChrysler. The plant itself is not their objective; just one of the many cargo containers that manufacturers like the automotive maker load with products for export. Everyday, Mannheim's main rail and terminal operator, Conliner, strings together miles-long freight trains for the fourteen-hour nonstop trip to one of the two Rail Service Centers in the Dutch port of Rotterdam. Mannheim and neighboring Ludwigshafen turn over sixteen million tons of cargo annually, making them the second largest inland trade center in Europe. Omar's colleagues have staked out the Conliner terminal for several weeks. Chainlink fences are along most of the perimeter, but security is modest because industrial shipments hold little allure for thieves.

One of their cyber-literate colleagues has hacked his way into the logistics database of one of the local manufacturers. He has downloaded the manifests and seal numbers for five containers destined for a well-established distributor in New York. The boxes are due to sail on a container ship steaming from Rotterdam to Port Elizabeth, New Jersey, in two days. The al Qaeda operatives monitor the rail shipments as they leave the factory and are delivered to the rail yard. After dark they select one of the boxes to carry their special cargo.

Gaining entry into the container is a matter of using wire

cutters to snip the flimsy fifty-cent lead seal that passes through the pad-eyes on the center edge of the two doors. As a bit of added caution, they have brought along a spare seal to replace the one they have cut. It's almost certainly overkill, since no one is likely to inspect the seal until the shipment is released from its port of arrival—and then it will be too late.

The terrorists mount their makeshift bomb on the floor of the container. The bomb has a triggering device that can be detonated remotely with a radio signal. In the event that someone tampers with the box before it arrives in the United States, the doors are booby-trapped to set the bomb off if they are opened. Since their preferred goal is to set off the bomb in New Jersey after it is unloaded from the ship, they need to be able to track where the container is. To accomplish this, they place a battery-powered global positioning system (GPS) transponder next to the bomb. The 7.5-inch, 27-ounce device will be able to signal, via a 3-inch external antenna, the container's location. The antenna that has to see the positioning satellites is affixed to the outside edge of the right-hand door with a pair of magnets, just above eye level. Since it is a wireless device—the GPS unit communicates with the antenna using a "Bluetooth" chip—it is hardly noticeable.

The container, along with hundreds of others, is loaded the next morning on a rail chassis and begins its half-day trip to Rotterdam. Rotterdam is Europe's biggest container port, and the largest seaport, by volume, in the world, moving 322-million metric tons of cargo in 2002. As a result of a new initiative begun by the U.S. Bureau of Customs and Border Protection in late 2002, there is a team of U.S. customs inspectors in the port of Rotterdam whose job is to identify high-risk containers that are destined for the United States. Working with their Dutch counterparts, roughly three dozen of the nine thousand outbound boxes leaving the port on any given day are scanned by a massive X-ray device that creates a digital image of the container and its contents. A well-trained technician could readily spot items like explosives attached to the container floor, but there is little chance that a box from an established German manufacturer will be selected for examination. Such a box would be moved directly to a pier with a fully automated stevedore process that uses machinery that remotely transports the container dockside. This means there is no risk that a curious longshoreman will spot anything untoward, since the only human being working in the terminal is sitting more than 120 feet above the box. The gantry crane operator snatches the container and loads it onboard a 910 foot container ship. Within ninety seconds the box is nestled in the hold of the ship for the ten-day transatlantic voyage, where it becomes nearly inaccessible. The vessel is manned by a crew of just fifteen, none of whom have the inclination, time, or ability in their shipboard routine to look closely at the 2,150 forty-foot containers in their custody.

While the dirty bomb is making its way to the Port of New York and New Jersey, a second cell of al Qaeda operatives has been working assiduously in Detroit, Michigan. Their orders are to launch an additional dirty bomb attack, timed to go off simultaneously in Detroit, and to provide the radioactive material to two other cells planning attacks in Los Angeles and Miami. The ring leader of the operation and head of the Detroit terrorist cell is an Algerian national—let's say his name is Hassan.

Hassan left Algeria in 1993. His departure was inspired by a crackdown on the fundamentalist militant organization to which he belonged. He applied for and received a one-month visa to France to visit a family member in Paris, but shortly after his plane landed, he traveled instead to Hamburg to take up residence with a fellow refugee. There Hassan was recruited into al Qaeda at a local mosque. He kept out of sight of the local authorities, using a phony French passport, purchased from a smuggler, for personal identification. In 1998, he volunteered to travel to North America to set up a local terrorist cell and flew to Canada. When he arrived in Montreal, a Canadian immigration officer discovered the fake passport. Hassan claimed he was a political refugee and asked to be given asylum status. Under Canada's liberal asylum policy, he was freed on his own recognizance and directed to appear for a hearing in six months. Hassan never showed up. He used the intervening time to procure a real Canadian passport that he applied for using forged documents.

Given his Algerian roots, Montreal proves to be the ideal base for Hassan since it is a cosmopolitan French-speaking city just an hour's drive from the U.S. border. The fact that organized crime has a substantial presence in the port, airport, and trucking industry is also helpful. Hassan's instructions from his al Qaeda bosses are to locate medical and industrial radiological materials for his dirty truck-bombs. The radioactivity of cesium-137 devices such as those found in high dose brachytherapy units or portable X-ray weld-inspection devices is relatively low and will not produce sudden death to the unlucky people near the blast environment. The attraction is their availability. The U.S. and Canada have over two million licensed sources of radiological materials, 500,000 of which are near or beyond the end of their service life. Three hundred sources of radiation have been reported as "lost" or "stolen" on an annual basis since 1996, 56 percent of which have never been recovered. By the end of 2001, there were five thousand "orphaned" radiation sources in the United States alone.

Over the course of five years, Hassan gathered enough cesium-137 to sprinkle in three fertilizer truck-bombs to be confident that the blast area will be contaminated with radiation. Shortly after Omar obtained the Azerbaijani amercium, Hassan received orders to smuggle the cesium into the United States and to rendezvous with a cell of Detroit-based operatives. Hassan decided that the safest route into the United States is through the Regis Mohawk Indian Reservation that straddles the Ontario– Quebec–New York border along the St. Lawrence River. The tribe owns over 30,000 acres of land, more than 14,000 of which are on the U.S. side of the border between Franklin and St. Lawrence Counties. The responsibility for patrolling this territory lies with a small tribal police force. But law and order has never really been a priority. Back in the days of Prohibition, the reservation was a magnet for bootleggers. Today, the contraband of choice is cigarettes, illegal migrants, and narcotics.

Pleasure vessels dot the St. Lawrence River in the hundreds on a warm summer day, so no one paid much attention to Hassan beaching his Boston Whaler on the southern shore, a short distance from a dirt road on the American side of the reservation. He grabbed his backpack and made the short hike to the waiting SUV. Two hours later, he was just outside of Syracuse, heading west on Interstate 90 for Detroit.

Detroit is home to the largest population of people of Arab descent outside of the Middle East. Hassan readily blends into the community there. He instructs the Detroit cell members to procure an unmarked white box truck and meticulously builds the bomb inside the van. Eight thousand pounds of ammonium nitrate fertilizer, mixed with the proper measure of diesel fuel oil and detonated with a TNT booster, will produce an explosion twice the size of the bomb that destroyed the Murrah Federal Building in Oklahoma City on April 19, 1995.

In the intervening days, he makes the handover of radioactive material to operatives sent to Detroit by the Los Angeles cell. Once the West Coast cell members return home, they go to work constructing their bomb in an abandoned warehouse in the Los Angles harbor community of San Pedro.

The Miami cell also sends a team to pick up the radioactive material. Their target is a large industrial park that serves as a free trade zone (FTZ) for freight that is using Miami as a stopping point on its way to the Caribbean and Central and South America. Every day, one in every three boxes that arrives in the Port of Miami is shipped fourteen miles inland to the FTZ, where the loads are broken down and the containers are repacked to meet the lower-quantity orders of these smaller markets. Because the goods are not actually being imported into the United States, there is very little record-keeping associated with these shipments. There are no duties to be collected so the entry paperwork receives only cursory treatment by customs inspectors.

One member in the Florida terrorist cell started work a year ago, driving a rig for one of the many Miami-based trucking companies that hire nonunion drivers. Every day, these rigs make up to six container runs from the port to the FTZ. The cell has purchased an old maritime container and a truck chassis to carry it on and stored it in the backyard of a rented home a few blocks from the FTZ. When the team arrives back from Detroit, they build their fertilizer bomb in the container. The plan is for the driver to swap the chassis with his first morning load for the one carrying the container with the bomb. The cell is particularly proud of the scheme to disguise the bomb in a FTZ-bound maritime container. Their aspiration is to create nationwide fear that there may be multiple containers being shipped to the United States that are armed with weapons of mass destruction.

The Detroit cell has a local job. Their mission is to target the world's busiest commercial border crossing by driving their truck on to the Ambassador Bridge that links Detroit, Michigan, with Windsor, Ontario. Each day, the bridge, which is owned by a private company, handles an average of 19,300 passenger cars and 9,100 trucks carrying a quarter of a billion dollars in trade merchandise. Hassan's hope is that the attack will lead to the closing of the entire U.S.–Canadian border, thereby severing the world's biggest trade relationship.

There is practically no risk of being intercepted getting on to the bridge, since the U.S. and Canadian border inspectors examine drivers and vehicles *after* they have crossed the Detroit River. The plan of attack will be to detonate their bomb as they pass under the southbound tower in an attempt to topple the column and cause the roadway to collapse.

On the early afternoon of September 3, the container ship from Rotterdam arrives at a marine terminal in Port Elizabeth. None of its containers has been targeted for examination by the customs inspectors at the Newark Field Office because the boxes have been effectively pre-cleared in the Netherlands by U.S. agents. After finalizing plans with the Detroit cell, Hassan drives to Newark with the detonator and GPS tracker the German cell smuggled to him in Detroit. He parks his SUV in the Elizabeth Center shopping mall parking lot adjacent to the port. From his parking space near a large Toys "Я" Us, he can look out on the most concentrated piece of transportation infrastructure in the United States—planes, trains, automobiles, and ships all sharing virtually the same real estate. It is the ideal place to cause mass disruption. If things go as planned, Hassan anticipates that the transportation lifeline that supports forty million people in a 200-mile radius will soon be severed.

With his GPS reader, Hassan can track the container's movement, first off the ship by the giant gantry crane, then to its assigned temporary storage place on the terminal by a small mobile crane known as a "straddler." The box will not be picked up until the following day, which is perfect for Hassan's plan to set off the explosive device precisely at 8:51 A.M., the time when the first plane collided into the north tower of the World Trade Center on September 11. He calls the teams in Miami, Detroit, and Los Angeles and tells them "the wedding is on" which is the pre-arranged signal to move into execution stage.

When the day breaks on September 4, it brings two unexpected bonuses for the terrorists. First, the winds are blowing ten mph from the southeast—ideal for carrying the radioactive plume out of the terminal, across the adjacent railroad tracks and the New Jersey Turnpike, and into the airport. Second, a local news crew has arrived to do a report on the customs container inspection operations. The port had just received its fourth large X-ray scanner with state-of-the-art software upgrades. The customs field office had agreed to provide a demonstration for the local New York network news affiliate.

Meanwhile, the Los Angeles team had scouted its target—a

huge electrical transformer station on Terminal Island that provides power to the port. The transformer is just off the roadway, surrounded only by a six-foot-high chainlink fence. It will be a simple matter to drive the panel truck through the fence, hop out into their getaway car, and detonate the explosives. The driver for the Miami team is at the port just as the morning shift starts. After he picks up his container, he heads to the rented house to make the swap with the container loaded full of explosives.

At precisely 8:51 A.M., Hassan presses the remote detonation switch and a massive explosion shoots a huge fireball into the air. He doesn't hang around to see what happens next. The explosion shatters windows in a nearby office building that has served as home for the customs inspectors since the late 1990s. The local media team immediately starts to capture the images of the bomb's explosive force. Then the alarms on the personal radiation detectors worn by customs inspectors start to go off. The agents have all been issued these devices in the aftermath of 9/11. The senior inspector orders everyone to take cover inside, but not before the news team mistakenly reports that a nuclear bomb has just gone off in the port. All the major networks break into their programming and begin to follow the story live.

There is soon much to report. When the container bomb goes off, its explosive force is felt by the morning commuters on the New Jersey Turnpike. The combined surprise over the noise and the intensity of the blast and the rubbernecking has the inevitable result—the kind of multicar accident that happens when highway drivers hit an unexpected fog bank.

Meanwhile, the air traffic controllers feel the shock of the blast and look out in the direction of the port from their tower to see the plume of smoke blowing toward the airfield. The senior controller orders the immediate grounding of all the jet airliners taxiing on the runway and directs the controllers to start diverting aircraft to other regional airports.

Commuters stranded behind the smashed up vehicles ahead of them tune in to their radios to hear news reports relay the mistaken story of a nuclear weapon explosion. They start bailing out of their cars in droves, fleeing as fast as their legs can carry them.

At the same time, on the West Coast, the 9.5 million residents of Los Angeles and Long Beach are just waking up and tuning in to the morning news as they wait for their coffee to brew. The white panel truck drives through the chainlink fence and backs up against the massive electrical transformer. Three minutes later, the truck disappears in a concussive ball of flame as the escape car speeds along the freeway out of Long Beach. People near the port thought of an earthquake, just for a moment, when the ground shook.

Two minutes later, the Miami team drives into the free trade zone. After showing his paperwork to the guard and being cleared through the gate, the driver pulls the truck just inside of the terminal, hops out of the cab, and sprints back out of the gate. The perplexed security guard is yelling at him when he is knocked down by the third explosion of the morning.

Detroit is host to a similar scene. Rush hour is well underway when the truck explodes on the Ambassador Bridge. Drivers and passengers in automobiles and tractor trailers in front of and behind the truck have no chance to escape as their fuel tanks erupt in flames. Images of the conflagration and damaged bridge are captured by a traffic helicopter and are broadcast nationwide. Then comes news that radiation has been detected by a customs inspector who rushed to the scene to try and help.

On the West Coast, the Coast Guard captain of the port of Los Angeles and Long Beach announces that radiation has been detected in the vicinity of the Terminal Island explosion. A few minutes later, a Miami fire department captain reports that his firefighters have picked up evidence of radioactivity while responding to the mysterious explosion in North Miami's free trade zone. He has ordered his men to evacuate from the scene until they can get the proper protective gear to fight the fire. The police and fire departments begin evacuating all the homes downwind of the explosion site.

At 9:55 A.M., the mayor of New York City announces that he has ordered all the bridges and tunnels into Manhattan shut down until further notice. This announcement is followed by news that the New York Stock Exchange and the NASDAQ have ordered a cessation in trading. Thirty minutes later, after a hurried meeting at the White House, the secretary of Homeland Security appears before the news cameras and announces the Homeland Security advisory system is being set at its highest level—condition red. He declares that for at least the next twenty-four hours, all border crossings and seaports will be closed to inbound traffic. Given its ongoing concern about the security of air cargo, the Transportation Security Administration orders all airports closed down as well.

Shortly after noon, a taped recording is played on the Arab Al Jazeera news channel. The voice of an al Qaeda leader claims responsibility for the four attacks. He also declares that there are more in store. He asserts that al Qaeda has three nuclear devices more deadly than those used that morning located in containers around the United States. Al Qaeda threatens to set off these weapons at a place and time of their choosing unless all the

American troops currently stationed in the Middle East are immediately removed from Arab lands and U.S. warships operating in the Persian Gulf and Red Sea depart from the region.

Around the United States, mayors and governors issue orders that trains and trucks will not be allowed into urban areas until they have been searched. The interstate highway system becomes littered with trucks stranded in breakdown lanes and rest areas. The national rail system grinds to a halt. The presidents of the International Longshore and Warehouse Union, and the International Longshoremen's Association join the Teamsters' president in issuing a press release in which they assert that the attacks highlight the extent to which the lives of workers are threatened by the failure of Washington to fund the backlog of port and transportation security measures. The unions declare that the longshoremen and truck drivers will not return to their jobs until these safeguards are put in place.

Getting the Port of Los Angeles up and running again is no simple task. The truck bomb completely destroyed the electrical transformer, and energy company officials are struggling to come up with a plan to restore power to the northern half of the port. The transformers, which are produced in Asia, typically require a two-year lead time to build and ship, and the U.S. electrical industry maintains no spares. The explosive force of the bomb and the prevailing winds spread radiation over much of the massive port complex and the surrounding neighborhoods. The oil terminal located on the south side of the harbor has been abandoned by the dockworkers. That terminal handles one half of all of California's crude oil imports. Because of the combination of high demand and limited fuel storage capacity, gas stations throughout the region will likely go dry within a week if the terminal is not reopened.

A preliminary survey of the Ambassador Bridge concludes that there is radiation contamination in Detroit and Windsor within a mile of the bridge. The explosion destroyed much of the roadway and caused significant structural damage to the suspension system and southern tower.

U.S.-bound traffic at all the major border crossings with Canada and Mexico is backed up for miles. In a conference call with the White House Chief of Staff, the CEOs of Ford, General Motors, and DaimlerChrysler advise that within twenty-four hours they will have to shut down their assembly plants on both sides of the border due to parts shortages. They estimate the loss of production capacity to be one million dollars per hour per plant. Tens of thousands of autoworkers will be on the unemployment rolls within a week.

Emergency responders in Los Angeles, Detroit, and Miami are having a difficult time keeping public anxiety in check because they do not have enough radiation detection equipment and protective clothing to provide an accurate report of how far the radioactive material has spread from the bomb sites. In the face of sensationalist media reports, the majority of the cities' populations decide to flee, jamming the roadways.

At the White House, the president instructs his Homeland Security Council that he wants a plan to search all the containers, trucks, trains, and ships that were inside the United States on the day of the attack. He is mortified to learn that this task will likely take several months. The Secretary of the Treasury brings more bad news: unless the transportation system is returned to normal operation within two weeks, virtually all the nation's major manufacturing plants and retail outlets will be shut down due to inventory shortfalls. Wall Street is paralyzed amid fear of a massive sell-off if the market is reopened. The overseas markets are in a free fall. The Secretary of Transportation pipes up that for every day U.S. ports stay closed there will be at least a week's worth of work to clean up the resulting congestion. A three-week shutdown will bring the global container industry to its knees.

A call from the governor of Hawaii makes clear how important it is to get the transportation system back up and running again. The islands only have a thirty-day food and energy supply. Even a Berlin-style airlift would not make a dent in meeting the state's cargo needs. The problem is being made worse by a public run on staple items—a situation that is starting to gather steam on the mainland as well.

Turning the system back on is not easy. Frontline inspectors are only in a position to physically examine 4 to 5 percent of incoming containers. Even with the mobilization of the National Guard to augment the inspection process, there is not enough space or time to look at everything and keep up with the inbound flow.

The Secretary of Homeland Security sums things up this way: "Mr. President, we have two options. The first is to open our borders and turn the transportation system back on as quickly as possible and hope al Qaeda is bluffing about launching follow-up attacks. The second is to play it safe and try to inspect everything. If we choose the first option and we are wrong, you will be blamed for placing a higher value on getting the economy running again than on protecting American lives, even in the face of a clear and present threat. If we choose the second option, we will cripple the nation."

All eyes turn to the president as he ponders the nightmarish decision that has to be made. Centuries before, the Roman Proconsul Quintilius Varus lost three veteran legions to the guerrilla tactics of Germanic tribes in the battle of the Teutobruger Forest. Legend has it that after his forces were decimated and before taking his own life, Varus was heard mumbling under his breath, "*ne cras*"—"not like yesterday." As the president weighs his next steps, he can only lament the fact that America had not spent its yesterdays preparing for the tomorrows that now confront the nation.

3

The Phony War

In September 1939, the Nazi army rolled across the Polish border and unleashed a new form of combat known as "blitzkrieg." The Polish army was routed, and the British and French governments finally abandoned their policies of appeasement and declared war on Germany. The next eight months came to be known as the "Phony War," a name drawn from the title of a book written by a middle-aged draftee in the French army named Jean-Paul Sartre.

The Phony War was a period of false calm before the storm that left the Third Reich in control of much of Europe. For most civilians, day-to-day life remained largely unchanged. If you were in Paris in April of 1940, it was not very different from being in Paris in April of 1939, with the exception that there were more people moving about wearing uniforms.

The British and French high commands largely frittered this time away by preparing for the kind of trench warfare they had fought in World War I. Confident that they had a combined navy, army, and air force larger than the Germans, London and Paris thought they could discourage Hitler from adding to his territorial conquests by activating reserves and reinforcing defenses on the Maginot line, mounted cannons that stretched for 250 miles along the Franco–German border. What they did not do was alter their battle plans to respond to the new offensive warfare that the Germans had effectively used in Eastern Europe. In May 1940, they paid a heavy price for their complacency. After months of relative quiet, Hitler's Panzer units raced into the lowlands, circumventing the Maginot line, and Paris fell shortly thereafter. The British expeditionary forces narrowly escaped across the English Channel from Dunkirk with little more than the shirts on their backs.

When it comes to confronting terrorism, the United States is going through its own version of the Phony War. Despite the periodic raising of the terror-alert level, our marching orders as citizens are to keep shopping and traveling. Instead of mobilizing a defense against enemies who are intent on targeting innocent civilians and critical infrastructure, the U.S. government is placing its faith in familiar national security formulas. Washington is acting on the false premise that the terrorist threat can be contained by taking the battle to the enemy, in overseas efforts to isolate and topple rogue states, and by hunting down the al Qaeda leadership.

Each terror-free day in America lulls us further into a false sense of confidence. To be sure, there is now a new Department of Homeland Security and the Pentagon has created a Northern Command to oversee the armed forces' contribution to defending the homeland. But we have not pressed our elected leaders to provide an accounting of the concrete steps they are taking to lessen our vulnerabilities or respond effectively if the worst should happen. For its part, the federal government has not provided specifics, often suggesting that tight lips are necessary to avoid giving the terrorists any ideas. Most Americans take this operational security imperative at face value and presume that there is a lot going on behind the scenes which the government cannot disclose.

The reality is that our old national security dogs are having a difficult time learning new tricks. The Department of Defense is not busy dusting off contingency plans to protect the homeland, because there are none on the shelf. For decades, the national security establishment has not been in the business of protecting the territory of the United States at or within our borders. The U.S. Air Force has kept an eye on our air space for incoming missiles and bombers. Beyond that, the Department of Defense has embraced a "forward defense" approach in which our troops are primarily based, and trained to fight, overseas. Our naval ships maintain command of the high seas, not the shallow water near our coasts. This is not just a matter of preference. The Posse Comitatus Act, passed in the aftermath of the Civil War, sets legal limits on the role the military can play in policing within the United States. But while Congress has seen fit in recent years to make the act less restrictive—primarily to nudge the Pentagon into participating in the drug war—the armed services have shown a strong bias to steer clear of anything beyond preparing for and waging foreign wars. Senior officers reflexively protest that they are warriors, not cops, and have steadfastly resisted anything that looks like domestic law enforcement.

In the case of the U.S. Navy, until recently, this desire to stay out of the homeland defense business even applied to safeguarding its own fleet within U.S. ports. This fact was highlighted for me nearly two decades ago while I was serving as the commanding officer of a Coast Guard patrol boat, the 82-foot *Point Arena*, based in Norfolk, Virginia, and manned with a crew of ten. My primary job was rescuing mariners who got into harm's way in the seas off the mid-Atlantic states, an infamous piece of coastline known as the "Graveyard of the Atlantic." But in April 1986, I was directed to attend a meeting at the Navy's Atlantic Fleet headquarters in Norfolk.

Arriving at the appointed hour, I was quickly ushered into a conference room and directed to a seat at the far end of a long table already crowded with senior naval officers. After the meeting was called to order, an intelligence officer stood and said that the Navy was worried that the 567-foot guided-missile cruiser USS *Yorktown*, with a crew of 366 men, might be attacked by waterborne terrorists when it returned from its deployment in the Mediterranean. The *Yorktown* held the distinction of being the first naval combatant to launch a barrage from the Gulf of Sidra in a Reagan-era attempt to pummel Muammar al-Qaddafi of Libya for his role in supporting terrorist attacks on America's overseas interests. Now, *Yorktown* was a potential target for retribution.

As the briefing concluded, the U.S. Navy captain who chaired the meeting made it clear why I had been invited. The Navy was not in the business of protecting itself once it entered U.S. waters. That job fell to the Coast Guard. The captain wanted to know what my plan would be to protect the *Yorktown*, and he asked how close I would allow a small vessel to approach before I would fire my ship's weapons to stop it. I thought of my cutter's ammunition, neatly stowed and collecting dust in the ship's armory. Six months before this meeting, with Coast Guard ammunition reserves in desperately short supply, I had been issued a gunnery moratorium. Simply put, I had been ordered not to have my crew practice firing our two .50-caliber machine guns because there were no bullets to replace the ones expended in training. The *Point Arena*'s weapons had never been fired in anger during my two-year tour, and worse yet, due to regular crew rotation, only three of my sailors had seen my weapons fired at all.

"Captain," I replied, "I would fire on a small boat only after it launched an attack on the *Yorktown* and had then decided to pursue my patrol boat as I am moving at flank speed in the opposite direction. With all due respect, your war plans may have me listed as a counterterrorism asset, but I am neither trained nor equipped for that mission." I went on to point out that the Navy had a fleet of heavily armed special warfare boats sitting at a pier in nearby Little Creek, Virginia. Surely those vessels would offer a better defense than what I could muster.

By the universal scowls that greeted me, I quickly surmised that they had been expecting a more gung-ho response. Predictably enough, the protest of a Coast Guard junior officer was overruled. The special warfare vessels, I was told, belonged to Navy SEAL teams that were used only in overseas deployments. So the following morning, my patrol boat was underway with my "force-protection" fleet that consisted of the Coast Guard cutter *Chock*, a 65-foot Harbor tug with a maximum speed of 9 knots, and two 41-foot and one 30-foot utility boats. Happily for the U.S. Navy, the terrorists were not on hand that day to challenge us, and the *Yorktown* safely arrived at its pier to welcoming friends and families.

My sink-or-swim foray into the domestic counterterrorism mission highlighted a fact of national-security life that was reinforced throughout my twenty-year Coast Guard career and remains the abiding reality today: well-heeled Defense Department players like the U.S. Navy are willing to go to extraordinary lengths to avoid getting saddled with security mandates closer to our shores. This was not just a Cold War phenomenon. As recently as the year 2000, the Chief of Naval Operations ordered up a strategic assessment of the world's megaports that did not include any in the United States. When a Navy captain on assignment to the Council on Foreign Relations offered to correct this oversight by examining what is arguably the most important strategic commercial port in the world—the harbor shared by Los Angeles and Long Beach—the top Navy brass balked. The Navy, he was told, does not do domestic security.

But Washington has long known that agencies that do shoulder the domestic security burden, like the U.S. Coast Guard, lack the staffing, training, or equipment to do the job. The Coast Guard is charged with protecting 95,000 miles of shoreline and an "Exclusive Economic Zone" that extends two hundred miles offshore covering 3.36 million square miles, with a force about the same size as the New York police department, deployed on a fleet of vessels that are among the oldest of thirty-seven navies around the world. Serious engineering casualties among its ancient fleet of cutters and aircraft are routine. And while the Coast Guard was handed more to do throughout the 1990s, from interdicting drugs and migrants to patrolling dangerously depleted fishing grounds, on the eve of 9/11, its force was pared back to its lowest level since the mid-1960s. Meanwhile, plans to modernize its fleet were limping along with barely enough funding to keep the program on a twenty-five year replacement schedule.

The Customs Service has been no better off than the Coast

Guard. The number of customs inspectors assigned to policing the millions of tons of freight that enter our country each day has been relatively flat since the early 1990s, even though the volume of trade almost tripled during that same period and its law enforcement mission grew as well. Budget requests to replace ancient systems for processing electronic customs documents from the trade community were consistently shot down throughout the 1990s. Now these programs are receiving some new infusions of funding, but it will be years before they are up and running. Almost half the inspection booths on the U.S. border with Canada were unmanned in 2001 because of staffing shortages. Many inspection facilities still rely on extensive use of overtime pay and the support of National Guardsmen to manage the work flow. Prior to 9/11, none of the major seaports, such as New York, Charleston, or Seattle, had X-ray scanners located in the port to examine the contents of a container. In the ports of Los Angeles and Long Beach, the port inspection facility is still located nearly six miles from the port because many years ago, the U.S. government's landlord, the General Services Administration, decided the rent was too high to locate the facility near the waterfront.

Then there are agencies such as the Centers for Disease Control and Prevention (CDC), the National Institute of Allergy and Infectious Disease (NIAID), the Food and Drug Administration (FDA), the Animal and Plant Health Inspection Service (APHIS), and the Food Safety and Inspection Service (FSIS). These organizations collectively provide the federal workforce to detect, intercept, and respond to an event of bioterrorism, such as an anthrax or smallpox outbreak, or an attack on our food and water supply with such destructive agents as ricin, botulinium toxin, or foot and mouth disease. Throughout the 1990s, budget constraints, hiring freezes, and far more attractive opportunities in academia and the private sector led to the graying of this workforce. According to a study released in July 2003 by the Partnership for Public Service, half of the federal scientific and medical personnel who work in jobs that support the biodefense mission will be eligible for retirement over the next five years. But the federal government has yet to marshal a comprehensive effort to address this crisis.

Law enforcement agencies like the FBI continue to face such basic problems as field agents who lack internet access and the means to receive e-mail attachments. Immediately after September 11, the Bureau had to rely on overnight delivery services to get photographs of suspected al Qaeda operatives. The FBI's inability to enter the information age has been not just an issue of budget. The insular nature of the agency has historically made it not only painful to communicate with anyone outside the bureau, but among fellow employees within the agency as well. A much needed "virtual" case file system that would give agents a better chance to connect the dots in unfolding terrorist plots is being developed, but is still not up and running. Things are not much better within the offices of the ninety-two U.S. Attorneys throughout the nation. U.S. Attorneys are responsible for running Anti-Terrorism Advisory Committees at the local level and for prosecuting federal cases involving suspected terrorists. They have not been provided with any additional staff to operate these committees. Because their budgets for administrative support are so small, when suspected terrorists are brought in for questioning, the interrogations are rarely transcribed and entered into an electronic data base. None of these offices has adequate manpower to analyze the information it collects in preparing and prosecuting cases. The result is that any recognition of common linkages among cases that point to broader conspiracies or new dangers is likely to occur by happenstance, not practice.

Over the past decade, the Immigration and Naturalization Service (INS) fared better in terms of budget than the other frontline federal agencies, but the doubling of its budget from $1.5 billion to $3 billion did not begin to correct the decades of bureaucratic neglect, nor did it resolve the conflicting priorities embedded in U.S. immigration statutes. Before it was broken up and reassigned to the Department of Homeland Security, the INS was charged with examining the 400 million people who entered this country legally each year, enforcing immigration laws against the estimated seven million people who are in this country illegally, processing millions of residency and citizenship applications, overseeing the deportation of 150,000 people a year, and running dozens of detention centers. Now these same mandates must be accomplished, plus counterterrorism, with its missions and people and resources sprawled across the new Bureau of Immigration and Customs Enforcement, the Bureau of Customs and Border Protection, and the Bureau of Citizenship and Immigration.

The Border Patrol saw its staffing numbers increased in the nineties to respond to illegal migration concerns along the southwest border. However, personnel levels are still not nearly what they need to be to stem the flow of illegal aliens, never mind stopping determined terrorists. A University of Texas study completed in 1998 estimates that it would take 16,000 agents to create a minimal deterrent along the Mexican border. The northern border remains largely unprotected with fewer than one thousand agents responsible for covering its five thousand miles, much of which is forest and water.

It's not as though people who are in positions of authority in both political parties have been kept in the dark about just how broken-down these frontline federal agencies are, or how poorly calibrated our national security establishment is to provide homeland security. Throughout the 1990s, congressional hearings amply documented these problems. A series of blue-ribbon panels commissioned by Congress, the Department of Defense, and the White House drafted comprehensive reports outlining the extent of our vulnerabilities to acts of terror. In 1997, the congressionally mandated National Defense Panel concluded that the defense community must adapt to the emerging threat of asymmetric warfare directed at nonmilitary targets. That same year, the president's Commission on National Critical Infrastructure Protection concluded that urgent attention must be given to the security conditions of every major sector from pipelines to the Internet.

In 2000, the Gilmore Commission highlighted in its first report the general lack of readiness among state and local police, firefighters, and emergency medical professionals to manage mass-casualty terrorist events. That same year, the Interagency Commission on Crime and Security in U.S. seaports rated the security of our major commercial ports as ranging from poor to fair—and the commission was focused on the threat posed by criminals, not terrorists.

Then in January 2001, the U.S. Commission on National Security for the 21st Century produced the last of three reports after a multiyear examination of the global security environment. The primary finding was that the most likely threat to the United States would be a terrorist attack on U.S. soil, and that the U.S. government was not organized to deal with that threat. A major exercise held that year, known as TOPOFF, highlighted the decrepit state of our public health institutions and the absence of federal, state, and local protocols to manage an act of bioterror. Another biothreat exercise sponsored by the Center for Strategic and International Studies, called "Dark Winter," confirmed the same problem.

The writing was on the wall before 9/11, but few in Washington wanted to read it.

A year after the events of September 11, another blue-ribbon group concluded that the U.S. government had made little meaningful progress in securing the homeland. The task force was convened by the Council on Foreign Relations and was cochaired by former Senators Gary Hart and Warren Rudman. Among its findings were the following:

- 650,000 local and state police officials were operating in a virtual intelligence vacuum, without routine access to terrorist watch lists.
- While 50,000 federal screeners were being hired at the nation's airports to check passengers, Washington was still dithering over providing new resources to bolster the inspection of containers, ships, trucks, and trains that enter the United States each day, even though there was widespread acknowledgment that a weapon of mass destruction could well be hidden in their cargo.
- First responders—police, fire, emergency-medical-technician personnel—were not prepared to respond to a chemical or biological attack.
- The critical infrastructure for refining and distributing energy to support the daily lives of Americans remained largely unprotected against sabotage.

- Governors were counting on their National Guard units to respond to a terrorist attack, but those units did not have protection, detection, and other equipment tailored for complex urban environments; nor did they have the special training to provide civil support in the aftermath of a large-scale act of terror.

The Council report concluded:

> America remains dangerously unprepared to prevent and respond to a catastrophic attack on U.S. soil. In all likelihood, the next attack will result in even greater casualties and widespread disruption to American lives and the economy.

Not surprisingly, in marked contrast to the pre–9/11 security reports that garnered virtually no media attention, this task force report received page-one and editorial-page coverage in the national press, and reports on its findings were featured by most of the major television networks. Still, across the board, progress on the report's primary findings following its release has been glacial. For instance, another Council on Foreign Relations Task Force Report, released in June 2003, calculated that the federal government was short-changing by one-third the dollars required to meet even the most basic emergency responder needs. Among the issues confronting the people who we will turn to first in an emergency are these:

- On average, fire departments around the country only have enough radios to equip half the firefighters on a shift, and breathing apparatuses for only one-third. Only 10 percent of fire departments in the United States have the personnel and equipment to respond to a building collapse.

- Police departments in cities across the country do not have the protective gear to safely secure a site following an attack with a weapon of mass destruction.
- Public health laboratories in most states still lack the basic equipment and expertise needed to respond to a chemical and biological attack, and 75 percent of state laboratories report being overwhelmed by too many testing requests.
- Most cities do not have the necessary detection equipment to determine which kind of hazardous materials emergency responders may be facing.

To this sobering sketch of American insecurity, we can add one final disheartening note. The one area to which the federal government has been most visibly dedicating its attention and resources since September 11—aviation security—remains frighteningly incomplete. Things have certainly improved when it comes to screening passengers. Nonetheless, while the flying public is busy shedding shoes and bags at X-ray check-in points, the tons of air freight being loaded in the belly of most commercial airliners continues to fly the American skies virtually uninspected. The U.S. government has hired nearly 50,000 passenger and baggage screeners to work in the new Transportation Security Administration (TSA), but there are little more than one hundred federal employees involved with monitoring the handling of air cargo. As long as air cargo comes from a company that has been shipping terror-free for two years, the TSA assumes it's safe. This assumption is a shaky one, since terrorist organizations like al Qaeda are certainly capable of penetrating these companies either during the loading process or by intercepting the containers while they are being driven to the airport by low-wage truck drivers. Only in November 2003 did the TSA announce random inspections of air cargo. Given the limits on resources, and the resistance of the air cargo industry to anything that might result in delays, these inspections will be little more than a token effort.

How is it possible that so little has been done and is being done? The reasons range from the ideological to the practical, but collectively they have served as a potent source of inertia. For one thing, the years of neglect have been bipartisan. The Clinton Administration promptly filed most of its domestic security reports on the shelf. Even after 9/11, the general public shows little evidence that it is keen to break with its decades-long habit of going about each day unencumbered by worries about security. Mayors, governors, and corporate leaders continue to demonstrate an extraordinary degree of deference to Washington on security matters, even though the federal government is clearly struggling to provide guidance in areas in which it traditionally has had limited reach or experience.

Part of the problem is that Washington continues to treat domestic and national security as distinguishable from one another. There is no more dramatic illustration of our reluctance to acknowledge that the two have become intertwined than the sudden arrival in our daily lexicon of the term "homeland security." Why did we feel compelled to embrace this term to guide our response to enemies who target us within our borders versus beyond our borders? For virtually every other country on the planet, the term "national security" does double duty. First, this concept encompasses the protection of the nation. Second, if there is any power left over, it seeks to protect its interests beyond its shores. The United States is the only country in the world where national security has historically been just about this second task.

In the aftermath of the September 11 attacks, we could have decided right away that we needed to broaden our definition of national security to include a concerted federal effort to confront our security imperatives at home, not just abroad. But instead—and without any evidence of debate—we decided to preserve and even reinforce the foreign-domestic breach in our security efforts.

The homeland security mission got its first official lease on life while the rubble of the twin towers was still smoldering. On September 25, 2001, President Bush announced the creation of an Office of Homeland Security in the White House and named Pennsylvania Governor Tom Ridge to head it. The president's charter for the office literally ran out at our water's edge—Ridge's charge was to coordinate domestic security activities within the United States and out to the territorial sea—twelve nautical miles from our shores. President Bush also announced that he was going to create a Homeland Security Council to be supported by a different staff than the National Security Council. The Defense Department, intelligence agencies, and our foreign policy establishment would continue to focus primarily on their traditional roles, looking to manage threats far from our shores.

As something of an inevitable outgrowth of those early organizational decisions, we now have a homeland security strategy *and* a national security strategy. The former is focused on coordinating prevention, protection, and response activities within U.S. jurisdiction. Our national security plan sets its sight on shaping the global environment to advance U.S. interests, including going after terrorist organizations and states who give them aid and comfort. From a narrow bureaucratic perspective, the appeal of this dual-track approach to security is straightforward. It keeps participants in the traditional national security establishment in their comfort zone. It does not upset age-old strategies and tactics, and it fends off a potential run on national security dollars to bankroll domestic security measures. The homeland security budget accounts and the national security budget accounts are on completely separate tracks, which practically guarantees that there will be no debate over the trade-offs associated with investing in one versus the other.

With so much of the federal government's emphasis and resources being placed on combating terrorism overseas, we should be having serious discourse about what the new national security environment means for our thinking about federalism. Our current approach is for the federal government to deal with the distant game and leave the job of tackling the home game to state and local government and to the private entities that own and operate our society's critical infrastructure. President Bush effectively said as much to the National Governors Association in March 2002. As Governor Ed Rendell of Pennsylvania reported after that session, "President Bush was honest and frank. He told us there's no more money for anything. He said essentially, 'You're on your own.'"

There is an equity issue here that deserves consideration. For instance, over 40 percent of all containerized cargo that arrives in the harbor of Long Beach and Los Angeles is destined for the

American interior. Is it appropriate that the security of that harbor be shouldered primarily by Los Angeles County taxpayers? Or how about the privately owned Ambassador Bridge, across which so much U.S.–Canadian trade passes each day? Should it fall only to the private shareholders to secure the world's busiest commercial border crossing? Since all capitalist roads lead to New York City, does it make sense that protecting the infrastructure of the financial capital of the world is almost exclusively a local and private responsibility? As things stand right now, when it comes to security measures that fall within the jurisdictions of local, state, or private sectors, the buck stops outside Washington.

Washington's hands-off approach to funding new domestic security imperatives is consistent with the prevailing wisdom, found in many quarters, that holds—when it comes to government—that less is more. An activist federal role in homeland security presumably would require an investment in government capacity at all levels. While tax dollars for national defense—as traditionally defined—have enjoyed widespread support since the Reagan era, so too has the agenda for restraining domestic government spending and cutting taxes.

What is wrong with this picture? First, it fails to take into account that most states and cities are in their worst fiscal shape in fifty years, and few see any bright spots on the horizon. A Century Foundation study released in July 2003 assessed the concrete ways in which the states of Pennsylvania, Texas, Washington, and Wisconsin have adapted to their new homeland security imperative. It found little evidence that states and localities have significantly improved protections for their residents. Clearly, when California is confronting multibillion-dollar deficits, and Oregon is shrinking the public school year and trimming vital health services for the elderly, they are in no position to find the resources to invest in training and new equipment for first responders, or to make new capital investments to protect critical infrastructure. And unlike the federal government, deficit spending is generally not an option. Borrowing sufficient funds to get their local security house in order would set them on an almost certain path to insolvency.

Washington's hands-off approach extends to critical infrastructure protection as well. This is largely due to the concern that security mandates will raise the cost of doing business for the private sector. Unfortunately, without standards, or even the threat of standards, the private sector will not secure itself. In fact, in the absence of clearly defined and well-enforced security requirements, companies that invest in protective measures for the parts of the infrastructure that they own place themselves at a competitive disadvantage. Why? Because infrastructure security suffers from the "tragedy of the commons," a dilemma first described in 1833 by a mathematician named William Forster Lloyd.

Lloyd outlined the tragedy of the commons this way. Imagine there is a town pasture open to all. Each farmer will use the pasture for grazing as much livestock as he can. The tragedy sets in when there are more cattle grazing than the land can sustain. The rational farmer weighs his self-interest of maximizing the number of animals that graze on the free grass against the public benefit of his exercising self-restraint so that the pasture is not overgrazed. As a rational being, if he believes that other herdsmen will not restrain themselves, then it is illogical for him to also exercise self-restraint. This is because the field will still be overgrazed, but he will have deprived himself of the interim utility of earning greater income by having well-fed livestock. So each herdsman will end up trying to maximize his gain, even though in the end, all the herdsmen will suffer as a consequence. Therein is the tragedy. As the eminent biologist Garret Hardin summed it up thirty-five years ago, "Ruin is the destination toward which all men rush, each pursuing his own best interest in a society that believes in the freedom of the commons. Freedom in a commons brings ruin to all."

Applied to critical infrastructure security, the tragedy of the commons works like this. Security is not free. A company incurs costs when it invests in measures to protect the portion of infrastructure that it owns or controls. If a company does not believe other companies are willing or able to make a similar investment, then it faces the likelihood of losing market share while simply shifting the vulnerability elsewhere. When terrorists strike, the company will still suffer the disruptive consequences of an attack right alongside those who did nothing to prevent it. Those consequences are likely to include the cost of implementing new government requirements.

Take the case of the chemical industry. By and large, the chemical industry has a good safety record. Security is another matter. Operating on thin profit margins and faced with growing overseas competition, most companies have been reluctant to incur the additional costs associated with improving their security. Now let's imagine that the president of a chemical plant looks around his facility and gets squeamish about the many lapses in security he discovers. After a fitful night of sleeping, he wakes up and decides to independently invest in protective measures that raise the cost to his customers by fifty dollars per shipment. His competitor, who does not make that investment, will be able to attract business away from the security-conscious plant because his handling costs will not rise. Capable terrorists and criminals will target this lower-cost operation since it is an easier target. In the event of an incident, particularly a catastrophic one, one of two consequences is likely. First, government officials will not discriminate between the more security-conscious and the less security-conscious companies. All chemical plants are likely to be shut down while the authorities try to sort things out. Second, once the dust clears, there will be a mad scramble by elected and regulatory officials to impose new security requirements that could nullify the earlier investments made by the proactive plant owner. Faced with such a scenario, the most rational behavior of the nervous company president is to continue to toss and turn at night and focus on short-term profitability during the day.

The only way to prevent the tragedy of the commons is to compel all the private participants to abide by the same security requirements. When these standards are universal, their cost is equally borne across the sector. Either as tax payers or as consumers, we will ultimately end up bankrolling these measures, but what we will be paying for is insurance against the loss of innocent lives and a profound disruption to our society and the economy.

In short, undertaking what is required to address our newly revealed vulnerabilities requires that Americans acknowledge the inherent limits of laissez-faire. In the area of security, the market cannot tackle its vulnerabilities on its own.

While the imperative of an activist federal role is unsettling for some on the political right, things are not much better on the political left. The homeland security mission has made many liberals squeamish. First, there are those who are apprehensive that any stepped-up focus on domestic security will inevitably place cherished civil liberties at risk. They point to the many sordid instances in which the national security rationale was abused by authorities in both hot and cold wars. Accordingly, they wish to prevent the horror of the September 11 attacks from being used as fodder by what they perceive as would-be despots who seek to expand governmental power to intrude on the lives of its citizenry.

Another source of queasiness among the left is the potential price tag of the new homeland security imperative. It's not that liberals have any real heartache about government spending, but they recognize that they are operating from a defensive position. Given the finite resources available to the federal government for discretionary spending—resources that will inevitably shrink in the face of large projected federal deficits—dollars spent on security could contribute to the further erosion of social programs designed to support the neediest in our country, to bankroll government services, and to fund public works.

The case for confronting the new warfare ends up having no natural constituency in the United States. For both the left and the right, acknowledging the new security imperatives within our national borders creates ideological dissonance. The traditional national security establishment prefers a business-as-usual approach to one where they are compelled to acknowledge an environment of threat that makes its role a secondary one. Federal agencies that find themselves on the front lines are, in the main, too broken to meet the demands of their day-to-day nonsecurity mandates, never mind the threat posed by tenacious terrorists. At the state and local levels, governors, county commissioners, and mayors are too busy fighting for their political survival in the face of dire budget deficits to focus on the new security imperative that the federal government is keen to shove their way. The private sector is keeping its eye squarely fixed on the bottom line. For most U.S. companies, security investments in the absence of federal support or a government mandate are not in the cards.

If anything, the forces that have combined to place us in a state of phony war are only growing, since there is a diminishing incentive both inside and outside Washington to acknowledge vulnerabilities. Elected leaders are becoming increasingly reluctant to convene vulnerability assessments that will highlight the need for costly security measures. To do so would only add to the political liability risk, should the threat transpire and the public discover that officials knew but failed to act. For private sector leaders, documenting problems they are unable to independently correct creates a legal liability problem that general counsels are keen to avoid. No one wants a paper trail documenting security lapses that could come back to haunt a corporation.

The general public is complicit in all this by its failure to insist on an accounting of what is being done to confront the threat of terrorism. There has been surprisingly little public appetite for answers to the question, "How did it happen?" Even though there is a paucity of evidence that anything has materially changed, Americans seem willing to go about their daily lives simply assuming that they are protected. Like those World War II Frenchmen who spent nine months in their cafés, confident that their government's preparations for trench warfare would keep the Germans a long way from Paris, we remain blind to the new form of warfare that could take a devastating toll.

4

Security Maturity

When it comes to dealing with the new security agenda, Americans need to grow up. We cannot afford to act as though 9/11 was just a freak event. Nor can we expect our government to secure a permanent victory in a war on terrorism. Washington can no more assure that there won't be terrorist victims on U.S. territory than a mother can guarantee she will catch her child whenever he falls. Terrorism is simply too cheap, too available, and too tempting ever to be totally eradicated. We must have the maturity both to live with the risk of future attacks and to invest in reasonable measures to rein in that risk. In other words, the best we can do is to keep terrorism within manageable proportions.

Security measures must be reasonable, because they are not sustainable if they clash with the freedoms of the populace or work in direct opposition to the imperatives of the marketplace. Inevitably, draconian measures will be abandoned as soon as the sense of emergency passes. At the same time, operating a modern economy and sustaining civil society requires an environment in which people and markets can interact, confident that today, tomorrow, and the next day are likely to be pretty much free of violence. Accordingly, for any society that hopes to prosper and have its democratic values flourish, some measure of security is not optional.

The good news is that security, pursued intelligently, can be achieved in ways that are fully supportive of civil society and individual rights. In many instances, the means to advance security can even be tailored to reinforce the market. This may strike many Americans as counterintuitive, but when it comes to domestic security, we lack sufficient experience to place trust in our intuition. What little experience we have has almost always been in reaction to specific episodes of terror, like the first World Trade Center bombing in 1993 or the Oklahoma City bombing two years later.

Because of our event-driven approach to addressing vulnerabilities, Americans have failed to appreciate that security works best when it is integrated into the normal course of business. When security becomes a reactive enterprise, pursued only after threats become manifest, the effort ends up being costly, ugly, and largely ineffective. Parachuting security measures in when they had been previously overlooked or treated as a secondary consideration can even be counterproductive. This is because quick security fixes can act much like a Band-Aid over an infected wound; while the bandage creates the illusion of treating the injury, it can cover more serious, untreated problems just beneath the surface.

One way to help Americans take a more thoughtful and constructive approach to security is to reflect on how we have come to manage the *safety* imperative over the past century. While improving safety and improving security may seem like entirely different things, these agendas actually share a great deal in common. For one thing, when the safety agenda first got underway, it was met with the same kind of public ambivalence—in some cases, outright hostility—as the security agenda today. While most contemporary Americans do not question the value of incorporating safety measures into the workplace, home, and recreation elements of our lives, many of our forebears felt differently.

At the turn of the last century, the annual loss of lives and limbs to workplace accidents ran in the hundreds of thousands. But safety measures that would have protected common laborers were generally viewed by captains of industry as incurring burdensome costs that undermined profits. Calls for industrial safety standards were opposed on the grounds that they involved unneeded and unauthorized government intrusions into the marketplace. In some instances, regulations were resisted by those who claimed that they infringed on individual liberties. In more recent times, opponents of laws that mandate the use of motorcycle helmets still justify their position in terms of protecting personal freedoms.

Over time, Americans have come to view safety not as a government-imposed burden but as a valued necessity. As our society became more urbanized and technically complex, people began to appreciate the benefits of devising and enforcing rules that reduced risk or harm through human error or mechanical failures. Along with the changes in public attitudes, businesses came to recognize that there was a market case for making safety investments. Safer factories have higher worker productivity rates and lower insurance costs. Safer products make for happier customers and fewer lawsuits. So while automotive manufacturers, for example, were once bitter opponents of safety mandates, today many try to outdo each other, advertising how their cars beat the safety standards established by the government and exceed the safety records of their competitors.

Our mature safety agenda also provides some helpful reference points for how to pursue security without turning the U.S. into a national gated community. First, it reminds us that the exercise is all about managing risk. The idea that it might be possible for our government to provide absolute security is as fanciful as ensuring someone's absolute safety. A myopic focus on perfecting security or safety is nonsensical.

Safety is always just a means to an end. Take the case of automotive safety. Cars can be dangerous—over 40,000 Americans die on U.S. roadways each year. Still, most people need to drive because it is their primary means of transportation. Automotive safety is about taking steps that allow people to drive while managing the risks that might cause injuries or deaths. If eliminating the risk of automobile fatalities were the goal, we would simply ban people from driving cars.

In another category of risk, many of the ingredients mixed together in chemical plants are among the most deadly substances known to man, yet these plants manufacture products that have useful industrial or consumer applications. Industrial safety is about lowering the risk that people will be hurt while producing products we need.

Safety regimes are not just about protecting lives but also protecting critical systems or functions. The Coast Guard operates Vessel Traffic Services to monitor and direct ship movements within some of the busiest U.S. ports. It does this not only to help prevent ship collisions that could injure mariners and others in the vicinity; its objective is also to prevent accidents that might lead to a major oil spill or a ship sinking in a channel. Such accidents would threaten the marine environment and could also close the port for an extended period to clean up the spill or salvage the sunken ship. We value both the environment and maritime commerce. Accordingly, taxpayers invest in a vessel traffic system that helps reduce the chance that these public goods might be harmed by large ships making the wrong maneuvers in close proximity to each other.

We also act to mitigate the consequences, should our preventive measures fail. Safety considerations are no longer an afterthought, undertaken only in response to specific accidents. They have become an organic part of how we conduct our daily lives. We ask, "What if?" as new products are being designed, and incorporate safeguards at the outset to avoid the most likely problems.

The rationale for investing in security should follow the same logic. The difference is that security focuses on developing countermeasures against people who consciously set out to cause harm and spawn disruptive consequences. We invest in safety because we presume that events beyond human control, or because of human error, will occasionally create bad results. In the post–9/11 environment, we now must plan for the eventuality that bad people are intent on making bad things happen. An architect designing a highrise office building in 2004 should be thinking not just about features that address the risk of an earthquake or fire but also how the structure and its occupants might be endangered if it were targeted by a terrorist. She then should contemplate ways in which that risk might be managed, such as designing ventilation systems that are capable of containing an airborne biological agent, thereby preventing its entry into a lobby or office space.

The safety paradigm suggests three key lessons that can inform the approach we take to homeland security. First, Americans do not expect their lives to be risk-free. Second, managing risk works best if safeguards are integrated as an organic part of a sector's environment, and if they are dynamic in adapting to changes in that environment. Third, government plays an essential role in providing incentives and disincentives for people and industry to meet minimum standards. Bluntly stated, security won't happen by itself.

Our current approach to homeland security largely overlooks all three of these lessons. Since 9/11, Americans have been subjected to a debate over homeland security that looks like a ping-pong match. On one side are those who advocate that we pay any price and bear any burden in a single-minded pursuit for fail-safe security. On the other are those who resist a stepped-up emphasis on security, asserting that it costs too much, makes the government too big and intrusive, and will fail in the end to stop determined terrorists. Few are talking about bolstering security as a process in which we evaluate the costs and benefits of possible measures with an eye to reaching an end point where risks of serious harm are satisfactorily addressed.

We need look no further than the U.S. border with Mexico to see how a spasmodic approach to security can backfire. Throughout the 1990s, the U.S. government stepped up efforts to tighten control over the southwest border. Border patrol agents were recruited in record numbers and outfitted with the latest surveillance technologies. More customs and immigration officials were assigned to monitor the legitimate flow of trade and travelers at the border crossings. Gamma-ray machines capable of scanning the contents of fully loaded tractor-trailers and railroad freight cars were installed, inspired by a concern that the North American Free Trade Agreement (NAFTA) vision of a single continental market would set off a stampede of illicit drugs and illegal migrants into the United States. But our newfound zeal has generated a rash of unintended consequences that have actually made the border zone increasingly harder to police.

Consider the case of Laredo, Texas, the busiest commercial crossing on America's southwest border. In 2002, three million trucks passed through this once sleepy border town 150 miles from San Antonio, over twice as many as when NAFTA began in 1994. Many of the trucks are old, poorly maintained, and owned by mom-and-pop trucking companies. The drivers of short-haul rigs earn as little as seven to eight dollars per trip. The annual job turnover rate averages 300 percent. The reason why there are so few state-of-the-art tractor-trailers with veteran drivers is that large trucking companies avoid the border like the plague, because it makes no economic sense to have better-maintained, long-haul trucks waste time running the gauntlet of frontline agents.

To avoid the cost of delays at the border, trucks carrying intercontinental freight typically off-load their trailers at depots on the outskirts of town. In the case of southbound traffic, a short-haul truck is then contracted to move the freight to a customs broker, who will verify that the paperwork is in order and that duties have been collected. Then the broker will order another short-haul truck to come pick up the shipment to transport the freight to the border, where it is subject to inspection by customs and immigration agents. After clearing the border, it is then dropped at a depot in neighboring Nuevo Laredo. Finally, a long-haul truck will be contacted to pick up the load from this depot and carry it into the interior.

Seem inefficient? It is. And it is just the kind of environment in which organized crime flourishes. There is an old adage in the cargo security business: "Goods at rest are goods at risk." Bad guys gravitate to the points of friction in the transportation system, because that is where the opportunities are richest to put in contraband or take out goods. Security professionals also know that poorly paid short-haul drivers, with high job turnover rates, provide ripe enlistment prospects for participation in criminal conspiracies. Since the small companies that own these trucks are operating on such thin profit margins, they are not in a position to perform background checks on their drivers, or maintain even the most basic security hardware, such as working locks for their cabs. But this vulnerable trucking sector would not exist if the border crossing did not create the market conditions that spawn and sustain it. If delays at the border were the exception instead of the rule, there would be little need for these brief, inefficient runs, and much of the short-haul trucking inventory would be making one-way trips to the junkyard.

A growing short-haul trucking industry that has become more difficult to police is not the only consequence of Washington's rush to harden the southwest border; it has been a windfall for professional smugglers in human labor as well. Stepped-up patrolling of the border may raise the risks associated with illegally entering the United States, but it also creates a demand for those who are in the business of managing that new risk by facilitating the illegal crossings. Migrants, who once simply strolled across the border to seek work on the other side, now need professional help. That help is provided by criminal networks of human smugglers known as "coyotes," who charge substantial fees for taking migrants to remote border locations or for transporting them through legitimate crossings. As the coyote business has become more lucrative, more money has been invested in building and refining these sophisticated smuggling operations. There is also plenty of cash with which to pay off Mexican frontline agents—and sometimes American ones as well—to look the other way. And it is not just desperate laborers who can use this service. As long as they are paid, the coyotes have no interest in asking why someone wants to cross the Rio Grande. Terrorist operatives keen to gain entry into the United States can afford their fees.

Our line-in-the-sand approach to enforcement also undermines the basis for cooperation and information-sharing among enforcement officials on both sides of the border. At a lunch in Laredo in mid-August 2001, I winced at the response by the U.S. Customs Service port director to my question about how U.S. and Mexican authorities coordinated their efforts to police the border. He said that it was tricky to talk with the Mexicans because they did not have secure telephones and one could never know who might be listening in. I pressed him, "Surely there must be instances where you need to coordinate enforcement operations involving criminals who operate on both sides of the border?" He acknowledged that such situations do periodically arise. When they do, he said, "I call my counterpart and ask him to send one of his men to the middle of the bridge. Then I put my message in a sealed manila envelope and send one of my agents out to the bridge as a courier." I found myself conjuring up an image from Hollywood spy movies where Soviet and American agents exchanged prisoners at Checkpoint Charlie.

This Laredo story illustrates why it is so important for security measures to be tailored to the dynamics of the environment they operate within. The rush to harden what historically has been essentially an open border has spawned an army of both legal and illegal intermediaries to work around the delays and hassles. Also, an environment that retards economic growth and integration will undermine the political foundation for closer collaboration among public authorities and the private sector on both sides of the border. The result is a border region that has become more difficult to police and one which terrorists could readily exploit to get their operatives into the United States.

Our handling of the southwest border over the past ten years offers a valuable lesson for improving security in the face of a terrorist threat: security solutions that focus only on the most visible manifestations of a threat will end up as short-term fixes that generate long-term problems. Happily, there is a way to avoid entering into a Faustian bargain that requires us to choose between pursuing shoddy cures or living with terrorists and criminals. The key is to identify opportunities in which security can reinforce the market's desire for efficiency and where it can support other valued public policy goals.

The necessary starting point for moving beyond the polemics that are now dominating this debate is to acknowledge that any single security measure always follows a curve of diminishing returns. That is, the harder you pursue a particular security fix, the more difficult it becomes to get much in the way of additional security results. Take the case of a door lock. The simple act of locking a door with a conventional lock will deter most amateur thieves. Since most thieves are indeed amateurs, this can be effective much of the time. Investing in a sturdier door and frame and a stronger lock may serve to foil more capable criminals, but it requires a far greater cost to get what ends up being only a small increment of added security. Those who are inclined to push the envelope in developing the "burglar-proof" door will end up generating the balloon effect. Bad guys will gravitate to the point of least security resistance. That is, the more difficult it becomes for a determined criminal to break in through the door, the greater the incentive for him to find another way into the house, such as through a window.

The way around these problems is to avoid the seduction of a silver-bullet approach to security. Anyone who claims they have developed *the* solution to a security challenge should be met with automatic skepticism. Effective security always requires constructing layers of measures. Each of these layers may be imperfect, but collectively they increase the odds of tripping up the bad guys.

Returning to our case of securing a home, we might consider some additional ways to ward off burglars without trying to make conventional doors and windows burglar-proof. Since thefts often occur at night, we could consider installing automatic lights that are triggered when people approach. A dog on the premises will provide another measure of security. Add to this a sign posted on the front lawn that indicates the home is monitored by a security company. Finally, the community could form neighborhood watch groups and post signs on the streets advertising this fact. Any of these measures independently might work only 60 percent of the time. But statistically, five 60-percent measures when placed in combination will raise the overall probability of preventing a burglary to 99 percent. In many instances, it may well be that the cost of all these measures is less expensive than trying to bolster any one or even two measures.

Our goal should not be to find fool-proof solutions for protecting the targets terrorists are most likely to strike. It is about identifying workable measures that are cost-effective and not disruptive. Then we need to string them together in such a way that each serves to reinforce the deterrent value of the other.

This is precisely the approach being undertaken at a place where the scars left by the World Trade Center attacks still run very deep. The two Boeing 767s that struck the World Trade Center towers on September 11, 2001, originated from Boston's Logan International Airport. That tragic fact still haunts the managers who work there, and the federal, state, and port authority officials at Logan have been quietly but aggressively changing the way they manage the risk of future terrorist acts. What they have achieved provides a template for how the nation should be tackling the terrorist challenge.

At Logan International, security has become everyone's business, from Craig Coy, the CEO of the Massachusetts Port Authority, on down. Each day at 8:30 A.M., representatives from forty different agencies, airlines, and service providers gather to attend a daily security briefing. The meeting is chaired by Thomas Kinton, the director of aviation. He is joined at the head table by George Naccara, the Transportation Security Administration's federal security director, who oversees the passenger and baggage screeners; Major Thomas Robbins, commanding officer of the Massachusetts State Police troop that patrols the airport, and Dennis Treece, head of Corporate Security for the Massachusetts Port Authority. The fact that the director of aviation runs the meeting highlights the emphasis that the airport's top management now assigns to security.

That airline and terminal officials participate in the daily security briefing might not seem particularly noteworthy. Unfortunately, it represents a striking departure from the norm. Traditionally, security people stick to themselves, fearful that information they share with anyone not directly under their control might end up in the wrong hands. Meanwhile, without a specific threat, operators see themselves as too busy to spend time worrying about security. The result is a wide chasm between those who run an airport and those who have the full-time job of trying to secure it.

In the case of Logan, closing the gap between the doers and the securers by fostering frequent interactions has yielded benefits beyond everyone sharing the same security picture. It has also contributed to the successful development of a number of impressive security innovations. With a "never again" sense of mission, the Logan Airport community essentially has taken a no-holds-barred look at how to rewrite the book on airport security without breaking the bank or generating terminal gridlock.

They began by recognizing that there are inherent limits to focusing primarily on passenger and baggage screening. What security officials needed were opportunities to intercept the bad guys before they got to the screening stations. They also had to close off the other ways in which terrorists might gain access to the airplanes, such as penetrating the perimeter of the airfield or mixing among the thousands of employees who have access to the aircraft. The result has been an innovative effort to build a series of concentric rings, each of which can help to elevate the probability of detection and interception of a terrorist threat. Collectively, these measures provide a powerful deterrent.

The airfield at Logan is in the heart of an urban area, and some portions of the property are abutted by residential homes. Much of the rest of it is surrounded by the busy waters of Boston

Harbor. A typical security measure is to attempt to create a *cordon sanitaire,* prohibiting anyone from entering a zone around the airfield. That is obviously tricky business when it involves private homeowners, and in Logan's case it would also have meant forbidding commercial fishermen from operating in waters where they had been harvesting clams for decades. The airport decided that rather than treating these neighbors as possible threats, it made better sense to enlist them as allies. The result was to create a kind of community watch where authorities met with the fishermen and property owners and outlined their concerns and how they could best help if they saw anything suspicious. The MassPort authority has gone so far as to hand out mobile phones to the clam diggers. Now that the airport has effectively deputized the eyes and ears of people who really know the neighborhood and have a vested interest in keeping it safe and secure, it has achieved the dual benefit of maintaining good community relations while multiplying its own security assets.

Of course, relying on an airport version of a citizen watch has its limits. If the neighbors are away or not paying much attention, someone can still find a way to sneak onto the airfield. Fortunately, technology can help to minimize the need for expensive perimeter patrols. Because airfields are always flat, open, and generally uninhabited by people beyond the area around the terminals, "smart" surveillance equipment with built-in software can be very effective. The equipment takes a digital image of the airfield and programs that image into remote cameras that monitor a given sector. These cameras can work in the dark and in bad weather. If something appears that differs from the stored image, an alarm is activated, highlighting the location where the anomaly occurred. The software can even be programmed to reduce the incidence of false alarms. It can be taught to distinguish the movement of a wild animal such as a bird or a deer from that of a person. When an alarm goes off, a watch officer then directs a security team to investigate.

Guarding the airfield perimeter is a relatively straightforward task. The more complicated security challenge is to identify and intercept a terrorist who seeks access to the airport by simply driving up to the terminal and entering the concourse. The authorities at Logan decided they would like to improve their odds of stopping an attacker before he gained access to the concourse area, where people board their planes. Their interest went beyond providing greater strategic depth. While hijacking a plane may be the endgame for some terrorists, others might be attracted to carrying out an attack on the airport itself. Setting off a high explosive curbside or in the terminal could generate mass casualties and disrupt a vital transportation lifeline that supports the local and regional economy.

The managers at Logan have sought to oversee the ground threat to the airport by periodically altering the traffic pattern of the airport, closing lanes and conducting random security stops. The idea is to prevent the airport from developing a predictable rhythm, because bad guys typically spend time staking out potential targets. The more a facility develops a routine, the easier it becomes to plan an attack. By making changes to the traffic flows, and the timing and intensity of security patrols, authorities are able to confound the kind of predictability that could inspire confidence among terrorists.

To protect people from a suicide bomber in the ticketing or baggage pickup areas of the terminal, and to enhance the odds of detecting someone who might be a potential hijacker, the state police troop that patrols the airport has been trained in "behavior pattern recognition" or BPR. BPR capitalizes on the simple reality that people who are about to commit suicide or engage in other acts of violence exhibit different kinds of behaviors than the everyday passenger trying to catch a plane. The state troopers have gone through an extensive training program run by Rafi Ron, the former head of security for the Ben Gurion Airport in Tel Aviv, Israel, to recognize these behaviors. If someone is acting in ways out of the ordinary, an officer is trained to approach them and ask them a few polite questions, such as where they are going, the purpose of their trip, and at what time their flight leaves. If the responses and the reaction are not appropriate, the person would then be detained for further questioning. To ensure that these techniques were not at odds with legitimate civil liberty issues, the program has been vetted and approved by the state district attorney's office and the American Civil Liberties Union. Now a scaled-down version of BPR training is being provided to the 15,000 employees who work at Logan under a program called "Front Line," and to the over 1,000 TSA employees who work in the airport under a program called "Logan Watch."

Terrorists might also get something dangerous on to an airplane by penetrating the army of airport workers that has access to the plane out on the tarmac. These workers range from food service providers, cleaning services, mechanics, and cargo handlers to contract construction crews. At Logan, this 15,000-person workforce uses employee entrances to access the secure area beyond the passenger checkpoint or to the airfield. To confront the risk that terrorists could pretend to be an authorized employee and use fraudulent identification to bypass passenger screening points, the airport has invested in an automatic scanning device to examine legal forms of identification. Employees are required to place their driver's licenses into the ATM-style reader and the device confirms whether the ID is genuine. The TSA inspectors also periodically set up checkpoint stations at these employee entrance areas, much like those in place for airline passengers. These random inspections help to ensure compliance with airport regulations governing who and what are allowed to be in the nonpublic areas of the airport.

The screening of passengers and their bags largely follows the familiar protocol now operating at most of the nation's 429 commercial airports, but Logan has sought to introduce several innovations. First is to provide the TSA checkpoint screening managers with a specialized version of the behavior pattern recognition, similar to what the state police have been receiving. These managers are assigned to work in front of the checkpoints as a kind of early alert system. They look for behavioral cues that warrant their screeners' taking special care in inspection of a traveler, or justify the traveler being referred for a law enforcement interview. Second, they are planning to install documentation verification devices at the passenger check-in points so identity checks can involve more than matching a photo to a face. Third, TSA deploys roaming teams to periodically reinspect passengers at gates. These inspections help to address the issue of prohibited items, such as carving knives or pizza cutters used in terminal restaurants, which might be obtained by a passenger after he has gone through the checkpoint. Also, if intelligence identifies that a terrorist has gained access to the terminal, it provides a backup means to intercept him.

Logan also took the exceptional path of developing an in-line system for inspecting unaccompanied bags. In most airports, explosive detection equipment is now located in the lobby of the terminal. After a passenger checks in at the ticket counter, she must carry her bags over to a machine for them to be examined. Once they are cleared, the bags are sent to be loaded on the airplane. The Massachusetts Port Authority made the decision to inspect the bags by the more efficient and less labor-intensive method of scanning them while they are sent on their way to an aircraft. Once a bag has been checked at the ticket counter or curbside, it moves by conveyor belt to a newly constructed facility where the luggage contents are examined automatically for explosive materials.

An important innovation that Logan's federal security director made was to put the TSA baggage inspectors through an intensive training program that allows them to ascertain quickly when the inspection machinery is generating false alarms. Inspecting bags of every shape and size that are filled with items that run from the eccentric to the everyday is an extremely demanding technical challenge. When the equipment cannot make a quick determination that something is benign, its default position is to assume that the item poses a threat. But rather than automatically tearing open the bags, skilled human operators can inspect them virtually and bring an additional level of sophistication to analyzing what triggered the alarm. In most instances, the operatives can quickly conclude that the bag does not pose a risk. This translates into important manpower savings and a reduced risk of delay and damage to passengers' property.

Officials at Logan have also taken a fresh look at the design for their newest terminal. By re-examining their pre–9/11 architectural drawings through a security lens, airport managers have discovered relatively low-cost ways to make the terminal easier to secure and to police while still optimizing efficient operations.

Should all these measures fail to stop a terrorist act, the airport has taken an important step toward preparing for the worst. In the winter of 2003, it staged a major disaster exercise at the airport that involved all the area hospitals, local emergency responders, and city and state officials. This kind of contingency will hopefully never have to be put into effect. Still, planning for how to cope with these eventualities can go a long way toward minimizing the chance that lives will be lost needlessly.

There are also breaches of security that turn out to be the result of carelessness, such as a passenger who belatedly remembers leaving something on a plane and rushes back through a terminal exit to get it. Without protocols for how to manage these kinds of events, the disruptive effect on airport operations can be the same as if the threat were real. For instance, in some airports, any breach of security automatically leads to an order to "dump the concourse," a response in which authorities order the rescreening of all the passengers who are waiting in the terminal or are onboard airplanes that have not yet departed. The resulting delays from holding up these passengers and their aircraft quickly ripple throughout the national air corridor, causing backups everywhere.

At Logan Airport, the top federal, state, and port authority officials have come together and worked out ways to rapidly assess breaches in security so that they can calibrate the airport's response. When a security incident takes place, the top leaders at the airport are immediately alerted. Officials then meet at or near the scene of the incident, or by phone, if they are not at the facility, to be briefed on the particulars. When available, they will review a videotape captured by a closed-circuit television camera. They will then determine whether there should be a partial shutdown of the airport, a complete shutdown, or a return to normal business. Since this system has been put in place, they have never had to order a full shutdown of the airport.

The example of Logan International Airport is instructive for all of us. Just as airport security clearly involves much more than screening passengers and their bags, our approach to homeland security has to be dynamic and multifaceted. Terrorists are in the driver's seat when selecting the time, place, and method of their attack. They will probe to identify our vulnerabilities and strive to evade or defeat our defenses. This means that security cannot be a sporadic investment made only in response to a specific threat or an actual act of terror. Instead, it must be constructed to operate with the same level of vigilance, nuance, and adaptability as a healthy human immune system.

Logan also provides a lesson on expanding responsibility for security to those not traditionally involved. In Logan's case, officials have reached out to their workforce—the airlines, concessionaires, and neighbors. For the nation at large, everyday citizens have to be an active part of the solution. We should draw upon a legacy that has been dormant since America mobilized to fight World War II. We should see it as a civic duty to participate in crafting and sustaining protective measures to ward off our enemies.

Finally, we need to plan for the eventuality that our security measures will be imperfect. In some cases, a security breach will be from a benign source. In others, attackers will succeed in carrying out their acts of terrorism. Whether an incident results in real casualties or is ultimately judged to be a false alarm, authorities at all levels need to know and rehearse what to do. As Logan's federal security director, George Naccara is fond of saying, "When you're in the midst of a crisis, you don't want to be exchanging business cards with the folks who need to help you out."

5

What's in the Box?

In March 2002, the group managing director of the world's largest terminal operator traveled from Hong Kong to Washington to find out how the U.S. government was handling container security and to extend an offer to help. John Meredith had spent much of his life at sea. At age fourteen he signed on to work aboard a British merchant ship. Two decades later, he came ashore and oversaw the construction of the first privately owned and operated marine container terminal. Now, as head of Hutchison Port Holdings, he sits on top of a $5 billion company that in 2003 moved nearly forty million boxes through its terminals in thirty-five ports in sixteen countries. While Hutchison owns no terminals in the United States, four out of every ten ocean containers that arrive in a U.S. port either originate from or pass through one of its facilities.

Meredith is one of the midwives of the global transportation revolution that has transformed trade. Before there were containers, the process of moving cargo between land and sea was slow and labor-intensive. Goods would be packed into a truck or rail-freight car, then unloaded at a seaport to be reloaded into the cargo holds of ships. When a ship arrived at a port, the cargo would be lifted out of the hold and transferred back onto a truck or train for the journey to its final destination.

With the advent of the container—they come in two sizes, but most measure 40' x 8' x 8'—and the creation of container ship terminals, all that changed. Moving upwards of thirty tons of cargo from a truck, train, or ship has become the transportation equivalent of connecting Lego blocks. The boxes can move across transportation conveyances without ever having to be opened, greatly improving efficiency. Large container ships can receive or discharge more than six million pounds of freight in a single hour. What once took several days now takes one or two work shifts. Today, the two largest container ports in the world, Hong Kong and Singapore, together handle more than a million forty-foot ocean containers each month.

It was during his trip to Washington six months after 9/11 that Meredith and I first met. I had drafted a proposal on how to improve container- and supply-chain security that was being circulated among the port communities in New York, Los Angeles, and Seattle. I thought Meredith would be like other transportation executives I had met whose focus was very much on the bottom line. Most of the captains of this industry seemed less concerned with another attack on U.S. soil than the prospect of bankrolling new security costs that would erode their already painfully thin profit margins.

Meredith is cut from a different cloth. He began our conversation by declaring that there was no doubt that containers are going to be exploited as a poor man's missile. The question is when, not if. Explosives, or even a weapon of mass destruction, could be readily loaded into a container at its point of origin or anywhere along the way to a marine terminal. Port terminal operators have no way of confirming whether what is advertised as the contents of a box is what is actually there. The measure of a commercial port's success is its ability to move cargo in and out of its turf as quickly as possible.

But that was not the only issue keeping Meredith awake at night. He was worried about the cascading consequences, should the United States decide to close its ports after a terrorist attack. He said that the industry had been able to cope with temporary port closures connected with natural disasters, such as a major storm or earthquake, by putting ships into a holding pattern and storing the outbound containers near the terminals until things returned to normal. But should the U.S. government close its ports for two to three weeks, Meredith warned, the entire system would go into gridlock.

On any given day around the world, more than 15 million containers are moving by vessel, truck, or train, or awaiting delivery. As megacontainer ships capable of carrying upwards of 3,000 forty-foot containers were put into operation in the 1990s, the need to choreograph the movement of boxes in and out of a marine terminal became more time-sensitive. With the explosive growth in trade over the past decade, many terminals are operating at or near capacity, twenty-four hours a day, seven days a week, which translates into the need for just-in-time delivery of outbound boxes, and quickly moving inbound containers out of the terminal gates. If the United States closed its ports to inbound ships, the result would be equivalent to a man tripping at the base of a crowded down escalator. To prevent him from being crushed by successive waves of arriving humanity, the escalator must be turned off. Similarly, port terminals must close their gates to incoming trains and trucks or else they will be buried under a mountain of containers that have no place to go. As a consequence, the trains and trucks carrying boxes to the port will be trapped outside the terminal gate. If they are carrying perishable freight, it will spoil and become worthless. But the more serious economic blow would be dealt to the manufacturing and retail sectors. Because 90 percent of the world's general cargo moves inside these boxes, when boxes stop moving, so do assembly lines, and shelves at retailers like Wal-Mart and Home Depot go bare.

Given what would likely be the catastrophic consequences of terror in a box, Meredith had been expecting someone from the U.S. government to contact him to discuss what Hutchison might do to aid in tackling the system's vulnerabilities. But in the three months following September 11, no one came. He spent the next three months trying to find out whom in the U.S. government he should approach with an offer to help. Having failed in that task from afar, he came to Washington to extend the offer to anyone who would accept it. After spending nearly a week making the rounds of the various government agencies, he concluded that while there were a few cooks in the kitchen, they all appeared to be working on different recipes.

Meredith had learned firsthand that the U.S. government simply was not organized to protect systems as critical to its national survival as the transportation system. The Coast Guard was focused principally on addressing the security associated with ships and their crews, but not the cargo they carried. The reach of the U.S. Customs Service extended to the cargo, but not the ships themselves or other transportation conveyances. With the Coast Guard assigned to the Department of Transportation (DOT) and the Customs Service, belonging to the Department of Treasury prior to March 2003, neither agency was working effectively with the other. At the U.S. Department of Transportation, the priority was to meet new legislative mandates to beef up security at airports. Even had it not been so preoccupied with airline security, the DOT would have had to overcome its largely domestic focus and compartmentalized approach, which treated the trucking, railroad, aviation, and maritime industries as distinct entities.

Presumably others in the federal government should be concerned about a security challenge that has the potential to bring the entire international trade system to its knees—the State Department, Treasury Department, Commerce Department, and the U.S. Trade Representative, for starters. Since this threat has obvious national security implications, it warrants high-level attention at the Pentagon, within the intelligence community, and at the Justice Department. But there is an appalling lack of engagement on this issue, despite the importance of global transportation to our national interests. For too long, port and container security has been viewed by these players as a backwater problem to be hashed out by technocrats and security professionals.

This tepid, piecemeal approach to container security is not exceptional. The situation is little better in the other vital sectors that support our daily lives, such as energy pipelines, power generation and distribution, information technology infrastructure, food and water supplies, public health, and toxic chemical production and transport. In all these areas, no single government entity has an uncontested charter to call all the security shots. Nor is there a standard by which to measure progress—or a lack thereof. Generally well-intentioned bureaucrats are left to their own devices to do what they can within their narrow scopes of authority. Not surprisingly, many of their initiatives are either redundant or in conflict with one another.

The lack of an accepted plan is not solely an issue of organization. It is also an outgrowth of both the public and private sectors' being intimidated by the magnitude of the challenge. It is one thing to marshal a plan to safeguard a government building or a national landmark. Even protecting a large airport like Boston's Logan International is daunting, yet strikes most as achievable. But how do you go about securing networks as vast and complex as the transportation system, energy pipelines, the electrical grid, communications, finance, health care, or food supplies?

Part of the challenge derives from the delicacy of the networks, which have become more sophisticated and interdependent in order to handle growing demand. James Woolsey, the former CIA director, has observed that complexity makes systems susceptible to a phenomenon known as the butterfly effect, whereby a small disturbance can produce unanticipated and profoundly disruptive effects across the network. On August 14, 2003, fifty million Americans found out how this can happen, when the lights went out in eight states, from Michigan to New York and into Canada. The chain of events that led to 263 power plants shutting down within a period of seven minutes began when three sagging power lines came into contact with the tops of overgrown trees in Ohio.

It is our total dependence on complex networks matched with their susceptibility to disastrous failures that make them such tempting targets for terrorists. But attempts to secure these networks might also generate the butterfly effect. In the aftermath of the 9/11 attacks, legislation was introduced in Congress requiring every container entering the United States to be unloaded and examined. It takes five agents three hours to completely inspect a fully loaded forty-foot cargo container. If that seems like a long time, think about how long it takes to empty the average moving truck carrying someone's household possessions. A container and an interstate moving truck are about the same size. On an average day, 18,000 containers are off-loaded in the ports of Los Angeles and Long Beach. If every box were unloaded and inspected, meeting the proposed 100-percent inspection mandate would translate into 270,000 man hours per day—which would require three times the customs inspection manpower that currently exists nationwide. Had this bill been enacted into law, the very process of trying to abide by it would have produced John Meredith's nightmare of global gridlock.

Another legislative approach was proposed in 2003 by Congressman Jerry Nadler, who represents a district within New York City. Nadler is rightly concerned that searching for a weapon of mass destruction in a container that has already arrived in a busy seaport would be too late, given that ports are usually near areas where people live and work. Nadler's solution, which he incorporated into a bill, is to require every U.S.-bound container ship to be boarded by the Coast Guard at least two hundred miles from our shores. From a landlubber's perspective, this might seem like a good idea, but there are daunting practical problems associated with routinely conducting lengthy inspections on the high seas. Not infrequently, the sea and wind conditions offshore make it too dangerous for the boarding team to climb up a ladder hanging over the side of the ship or to be lowered from a helicopter. Containers can be stacked up to eleven high and the space between them is often no more than eighteen inches—which makes gaining entry into the box impossible. If the process were anything beyond pro forma, the inspection of a single ship could take several days.

The inherent limitations of trying to inspect cargo at sea or within our ports suggest that we must radically rethink the bomb-in-a-box challenge. Securing cargo containers boils down to three things. First, there should be a system in place that ensures that only legitimate and authorized goods are loaded into a container. Second, once a container is on the move within the global transportation system, there should be measures that protect the shipment from being intercepted and compromised. Third, each port should have a rapid and effective means to inspect cargo containers that arouse concern. Once a box leaves a factory, it should not be open game for thieves to take items out or for terrorists to put weapons in. Inspections at borders should be about checking that these point-of-origin and in-transit controls have not been violated.

The challenge of securing the loading and movement of containers is formidable. Anyone who has $3,000 to $5,000 can lease one of the many millions of containers that circulate around the globe. They can pack it with up to 65,000 pounds of items, close the door, and lock it with a seal that costs a half-dollar. The box then enters the transportation system, with all the providers working diligently to get it where it needs to go as quickly as possible. Accompanying documents usually describe the contents of the cargo container in general terms. If the box moves through intermediate ports before it enters the United States, the container manifest typically indicates only the details known to the final transportation carrier. For instance, a container could start in Central Asia, travel to an interior port in Europe, move by train to the Netherlands, cross the Atlantic by ship to Canada, and then move by rail to Chicago. The manifest submitted to U.S. customs inspectors often will only say that the container is being shipped from Halifax and originated from Rotterdam.

If a container is destined for a city inside the U.S., only in exceptional circumstances would it be inspected at the arrival port. On any given day there are thousands of containers that arrive on the East and West Coasts that are loaded on trucks or trains to travel to the heartland of America as "in-bond" shipments. These containers have up to thirty days to get to Chicago or Pittsburgh, where the customs examiners in the destination port assume responsibility for releasing the container into the economy.

On average, overseas containers will pass through seventeen intermediate points before they arrive at their final U.S. destination, and often their contents come from several locations before they are even loaded into the box. Nearly 40 percent of all containers shipped to the United States are the maritime transportation equivalent of the back of a UPS van. Intermediaries known as consolidators gather together goods or packages from a variety of customers or even other intermediaries, and load them all into the container. Just like express carriers in the U.S., they only know what their customers tell them about what they are shipping.

Despite the complexity of this shipment process, the U.S. approach to monitoring the flow of boxes is startlingly simple. U.S. customs inspectors divide the universe of containers into two categories—trusted and untrusted. A trusted container is one being shipped to an importer or by a consolidator who is known to customs inspectors. Essentially, they are repeat customers who have no history of smuggling or trying to violate other U.S. laws. These boxes are cleared by customs officers without any examination. Untrusted containers are those that come from the world's trouble spots, from new importers who have no established record of clearing customs or who trigger some other alarm, suggesting that an inspection is warranted. Customs field inspectors are alerted to which containers they should treat as trusted and untrusted by the National Targeting Center, which evaluates information found on the cargo-container manifest and the customs declaration form and correlates it with intelligence. Based on a computerized Automated Targeting System, which assigns a score to each box based on risk, the National Targeting Center alerts customs inspectors in a port to hold selected boxes until they can be examined.

In theory, this approach is a sound one. Just as the Internal Revenue Service does not audit the returns of every taxpayer, it is foolish to incur the costs of opening every container in order to make sure that importers are not lying about the description, value, and quantity of what they are bringing into the U.S. The vast majority of companies are legitimate and law-abiding, and facilitating the movement of legal goods is important to our economy.

However, when it comes to counterterrorism and the fact that people's lives are at stake, the problem with the trusted-shipper approach is obvious. The stakes associated with mistakenly designating a container as low risk can be enormous. Rather than loading a weapon in a first-time shipment from a company in Afghanistan, which will almost certainly be selected for examination by U.S. inspectors, terrorist organizations are likely to take the time to figure out how to target the shipments of an established company. Current transportation and logistics practices provide fertile opportunities for groups like al Qaeda to compromise these legitimate shipments. In fact, in the post–9/11 world, we should assume that bad guys know who a trusted shipper is and will target a trusted box first. It follows that a top priority must be to move from the current "trust but don't verify" system to one where verifiable measures are in place to protect all shipments.

What should a new transportation security regime look like? The approach we need to take must be informed by several underlying principles. We have to recognize that the networks we rely on today are integrated into much larger continental and global systems. We can no more protect these critical infrastructures exclusively at home than a computer-security manager can successfully protect his network by focusing on the server next to his desk. Nor is it only our borders that need to be protected. Borders represent only a territorial line where a threat might enter into sovereign jurisdiction, but functionally the threat starts much farther back.

Next, we must constantly be mindful that the resulting state is not perfect security, but risk management. Risk management is partly about trying to reduce the probability that terrorists will succeed, and partly about reducing the likelihood that our response to an attack will cause more harm than the attack itself. It is the overreach in the aftermath of terrorism that makes it such an attractive means of modern warfare, and the likelihood of overreaction will rise inversely proportionate to the lack of credible safeguards developed in advance of the incident.

We need to understand that risk management is impossible if

there is not sufficient visibility and accountability in the network. Managing the risk of nuclear Armageddon during the Cold War involved trillions of dollars for monitoring the Soviet Union. This was done in large part to prevent inadvertently striking first. Moreover, transparency is essential if the public is to believe that the government is maintaining effective oversight of current security imperatives.

We also have to take as a given that we will rarely have explicit advanced intelligence about the nature of future attacks. Accordingly, we will have to look proactively at our critical infrastructure from the perspective of the terrorists. These vulnerability assessments must look at the entire system and its links to other systems, and not be limited to a specific component.

Providing security to complex systems must be a collaborative exercise. It is essential to place special emphasis on those initiatives that provide multiple benefits. By pursuing these measures with vigor, it is possible to build up the kind of trust that is necessary in confronting harder decisions that involve sacrifices.

Finally, we have to recognize that security is dynamic and requires multiple measures arranged in layers. Our enemies always will be probing our systems for weaknesses. A Maginot-line approach will work no better than it did for the French in keeping Nazi tanks out of Paris.

Applying these principles to the task of securing the system that moves millions of containers on any given day might seem Herculean, but it turns out that the problem is more manageable than the numbers suggest. This is because virtually all boxes will pass through just a handful of seaports if they are going to find their way to the United States. In fact, approximately 70 percent of the eight million containers that arrived in U.S. ports in 2002 originated from or moved through just four overseas terminal operators; Hutchison Port Holdings, P&O Ports, PSA Corporation, and Maersk–Sealand. That maritime transportation is concentrated in so few places and managed by so few hands makes it an extraordinary pressure point. The major terminal operators should be the gatekeepers who ensure that only secure boxes will be loaded on to ships that cross the Atlantic and Pacific Oceans. Their job would involve assisting authorities to accomplish two things. First, they should be able to help confirm that a low-risk container is in fact low-risk. Second, if a container has been deemed high-risk, it should be handled in a way that poses a minimal level of danger and disruption.

To guarantee that a container belonging to a trusted shipper has not in fact been compromised, we should insist that it be loaded in an approved secure facility at its point of origin. These facilities would have loading docks with safeguards that prevent workers or visitors from gaining unauthorized entry. The loading process would be monitored by camera. A digital series of photographs, each with a time signature, would record the interior of the container when it is empty, half full, and full. A final image would record when a security seal is activated, and all these images would be stored on a data chip with the container, or be transmitted electronically to the appropriate authorities in the loading port.

The container should be outfitted with light, temperature, or pressure sensors that could detect an unauthorized intrusion. Additionally, there should be an internal sensor that could detect indications of prohibited items such as gamma and neutron emissions associated with a nuclear weapon or dirty bomb, prohibited chemicals and biological substances, or CO_2 generated by a stowaway. A container-tracking device could keep a global positioning system (GPS) record of the route that the container travels. The truck drivers moving the container could be subjected to background checks. If a driver is going through areas known for smuggling or terrorist activities, a form of invisible-fence technology could be outfitted inside the truck. If the truck strayed from its designated route, a microcomputer would record the incident and later automatically idle the engine before the truck arrived at the port terminal. A radio signal would transmit an alarm to the relevant authorities, providing them with advance warning of the suspicious activity.

Once a container arrives at a terminal, it would have to pass through a nonintrusive inspection unit equipped to detect radiation, interrogate the sensors installed in the box, and create a CAT scan-style image of its contents. This image, along with other sensor data, would be forwarded through a secure Internet link to all the national customs authorities along the route. Sharing data records would allow experts to remotely look over the shoulders of frontline agents. Knowing that their inspection could be double-checked would make these agents less willing to accept a payoff to look the other way. This extra set of eyes would also provide another chance to detect problems. Even if the container is mistakenly allowed to be loaded on a ship by an overseas agent, the ship could be ordered to stay offshore until the container is inspected at sea.

Ensuring that a box could be found after it has been loaded and that it is not diverted from its advertised route means that authorities have to be able to track a ship once it has left a port. Most Americans would be surprised to learn that while civilian air-traffic controllers and the U.S. Air Force can track aircraft, there is no equivalent system for monitoring the movement of ships on the high seas. While creating this capability is not technically difficult, it has never been mandated. The closest we have come is a new requirement that large ships carry a device that allows the Coast Guard to detect them when they are twenty to thirty miles from our shores.

The lack of a tracking system has long been a source of frustration within the maritime law-enforcement community. I recall several drug smuggling cases in which an undercover Drug Enforcement Administration (DEA) agent would report witnessing a load of cocaine being hidden on a vessel. The agent would provide a description of the vessel and its departure time from a South American port. But unless there was a Coast Guard or Navy ship or aircraft nearby to locate and track that vessel—and there rarely was—the U.S. government couldn't find it. Law enforcement officers might realize three weeks later that the vessel had actually traveled to the Port of Philadelphia, discharged its cargo, and left the week before.

Assuming that a ship made it into port without incident, its containers should be selectively spot-checked. Containers should pass through radiation detectors, and a scanned image at the arrival port should be compared with the image taken at the loading port. If the images and sensor data match, it can be safely concluded that the shipment has not been tampered with and it can be released. The containers should then be tracked as they move to their final destination, allowing the ability to intercept the shipment in the face of late-breaking intelligence.

This level of attention should also apply to outbound U.S. cargo. Port terminals could be targeted by land as well as by sea.

A domestic-based terrorist could put a bomb into a shipment of exports, and then set off the explosive device once it arrives in the stateside port facility. Additionally, the U.S. needs to practice what it preaches if it wants to sustain support from its trade partners for their efforts to examine containers destined for American shores.

The challenge of policing our own ports is not as unmanageable as the volume of containers suggests. The top five maritime loading centers in the United States handle almost 60 percent of the containers exported from the U.S. Fifty percent of the containers that we export are actually empty, because Americans import much more than they export. Inspecting these empty boxes can be accomplished quickly and effectively. The remaining containers can be screened in the same way as those destined for the U.S.

This combination of harnessing new technologies and designing the means to check and double-check the status of shipments would help accomplish several things. It would create an effective deterrent against terrorists shipping a nuclear weapon in a container. Right now the odds stand at about 10 percent that our current targeting and inspection practices would detect a device similar to a Soviet nuclear warhead surrounded by shielding material. By using a mix of sensors and more vigorous monitoring, we could push the probability of detection into the 90-percent range. Given the difficulty of obtaining a nuclear weapon, a terrorist organization would think long and hard before taking on those kinds of odds.

Also, if authorities received specific intelligence that a weapon had been mixed in with a shipment destined for America, outfitting containers so they could be tracked would provide the means to act on that intelligence without disrupting the rest of the transportation system. Today, just the opposite would occur.

Consider this hypothetical situation. Imagine that the CIA has managed to infiltrate an al Qaeda cell operating in Karachi, Pakistan. One day, in a nondescript warehouse within that sprawling city, the agent witnesses a chemical weapon being loaded into a container bound for the United States. After he watches the truck pull away from the loading dock as it heads down to the port, he sneaks away to put in a call to the agency. The CIA watch officer notifies the White House Situation Room, and two hours later, the president meets with his national security and homeland security advisors. He turns to the Secretary of Homeland Security and asks for details on where the container is now and where it is headed. The secretary responds that the manifest indicates the box is headed to Seattle, but it could have moved by coastal freighter to any one of the major Asian ports. Once it has been loaded on an east-bound container ship, it could be heading to the ports of Vancouver, or Seattle, or San Francisco, or Los Angeles, or perhaps it is steaming toward the Panama Canal to be delivered to a port city on the Gulf or East Coasts. But he assures the president that all his inspectors will be on the lookout for it when it arrives. The Commander in Chief is not likely to be reassured. His only option might be to order all inbound container ships stopped in order to find a single weaponized container.

Of course, the kind of detailed intelligence in this scenario will almost certainly be in short supply. The only way to compensate for that is to establish sufficient visibility within the network, allowing a credible means to detect and intercept abnormal behavior. The computer industry does this to catch hackers. The cyber-security process involves mapping how traffic moves in the most technologically rational way. Once this baseline is established, software is written to detect aberrant traffic. A competent computer hacker will try to look as much as possible like a legitimate user, but because he is not legitimate, he inevitably must do some things differently. Good cyber-security software will detect that variation and deny access. For those hackers who manage to get through, their breach is identified and shared so that this abnormal behavior can be removed from the guidelines of what is normal and acceptable.

In much the same way, the overwhelming majority of global shipments travel in predictable patterns. If regulators and enforcement authorities use sophisticated data analysis to monitor those flows and the commercial documents associated with them, they can develop a comprehensive picture to enhance their odds of detecting abnormal behavior. Criminals and terrorists who seek to exploit legitimate commerce almost always do something out of the ordinary. This is because they have something to hide, and they usually lack the knowledge and experience to abide by all the rules of the marketplace.

Should detection and interception efforts fail, visibility also gives government authorities the means to quickly answer the question, "What went wrong?" If it takes days or weeks to determine just how an attack happened, every box will be viewed by a frightened public as another potential weapon. Supply-chain visibility can help in the same way that cockpit and flight data recorders are used in accidents involving passenger aircraft. Finding these recorders and providing an early indication of the probable cause of the accident play an important role in getting passengers back on planes after an airline disaster. Similarly, if government officials have the ability to quickly identify a bomb's origin, they would have a better chance of calming the public without having to shut down the entire transportation system to verify that it is free of explosives.

Developing the means to track and verify the status of containers provides benefits that go beyond security. There is a powerful commercial case to be made for constructing this capability as well. When retailers and manufacturers can monitor the status of all their orders, they can confidently reach out to a wider array of suppliers to provide them with what they need at the best price. They also can trim their overhead costs by reducing inventories with less risk that they will be left short.

Transportation providers will benefit from greater visibility as well. Terminal operators who have earlier and more detailed information about incoming goods can develop load plans for outbound vessels in advance and direct truck movements with greater efficiency. Companies like APL who own fleets of containers can optimize their use to a far greater extent than today. APL owns about 300,000 containers. When they are on a ship, the company knows where they are, but once they land, it does not. It only finds out when customers call to schedule a pick-up once they have emptied out the container. A typical delivery contract allows a company up to ten days to empty a container before incurring additional fees. Most containers are emptied within twenty-four to thirty-six hours, but companies often wait until the last minute to contact APL to come get the box so that it can be put back into circulation. Now imagine if containers had sensors that could indicate precisely when they are empty and send an alert message that includes the box's location to APL. William Hamlin, the man responsible for running APL's operations in North America, believes his company could start routinely recovering its containers within two to three days instead of the typical eight- to ten-day interval. As a result, a container used for just five full loads a year could be used for six instead, a 20 percent increase in productivity.

Greater visibility also brings potential benefits for dealing with insurance issues. Knowing precisely where and when a theft takes place makes it easier to decipher the nature of the threat and to identify what breaches, if any, contributed to the loss. When there is damage, it is much easier to track down the responsible parties. In short, rather than insurers spreading the risk across the entire transportation community, they can more carefully tailor insurance premiums. In turn, this creates a stronger market incentive for all the participants in the supply chain to exercise greater care.

Putting this comprehensive system in place to ensure end-to-end visibility and accountability of containerized cargo does not require futuristic technologies. Taking and transmitting digital images is now routinely done by proud parents who want to send baby pictures to distant friends and relatives. General Motors has its OnStar service; which allows it to find a car if it is stolen, to alert emergency personnel if the air bag is deployed, to remotely diagnose an engine problem, or to unlock a car if a customer leaves his key inside it. Sensors that can be built into a container are under development and will probably have a lifetime cost of around $250 per box, if widely deployed. To put that cost into perspective, the average container is used for ten years. That means that over the life of the container, the initial cost of installing sensor technologies into the box would add about $5 to the price tag of each shipment.

Radio frequency transceivers are now in common use across the northeast by commuters who use electronic toll systems such as E-Z Pass. These devices can store data that range from a single identification signature to thousands of records. Within the United States, virtually all railcars have these transceivers installed so railroad companies can provide their customers with ongoing position reports of where their freight is and when it will arrive at its final destination.

The latest radiation detection portals and container scanning equipment are being combined into a single unit and can capture images of trucks moving at speeds up to ten mph. These units cost about one million dollars each. Large ports would need several to ensure that the screening process would not slow the flow of trucks. They would also need to have spares on hand to allow for routine maintenance or to swap out a unit that breaks down for some reason. Developing a secure network to share and analyze the scanned images across multiple jurisdictions is a matter of investing in new command centers with strong information-technology backbones and well-trained analysts.

In the age of GPS, there is no technical barrier to tracking ships on the high seas. In fact, virtually every ocean-going vessel that travels the Atlantic and Pacific Oceans maintains regular contact with its parent company by using a system called Inmarsat. It is essentially a mobile phone that uses satellites instead of land-based antennas. Whenever an Inmarsat radio is used, the satellite knows the precise location of the caller.

Hutchison Port Holdings has already invested millions of dollars installing new equipment in its terminals to better monitor the location and integrity of containers in their custody. Given the company's role as a market leader, other modern terminals are likely to follow. The main barrier preventing terminal operators from installing new hardware and playing the gatekeeper is a concern that if the security bar is raised, shippers might take their business to smaller ports that provide less security and its associated costs. Port managers also have concerns about how a larger policing role might effect port operations. What if there are many false alarms? Even an alarm system that is 99 percent reliable could create real problems in ports like Los Angeles. There, a one percent false-alarm rate translates into 180 containers a day that would have to be investigated. Pulling that many boxes out of the queue can be expensive and time-consuming. The other serious issue is, what happens when the alarm goes off because of a real threat? If a dirty bomb is discovered, who will dispose of it? If anthrax is discovered, how is the shipment to be handled?

Offsetting these legitimate concerns is that the largest ports overseas understand how devastating the consequences would be if the transportation system had to be shut down even temporarily in the aftermath of an attack. Since the United States is their biggest market for outbound shipments, the ports also want to maintain good relations with U.S. authorities. Finally, most recognize that they make an attractive target. The costs of having an undetected weapon of mass destruction go off at a facility would be incalculable both in human and economic terms.

To have the incentives for action trump the incentives for inaction, we need to establish "green lanes" in seaports. The concept is essentially the same as the E-Z Pass toll collection system. The reason why commuters love E-Z Pass is that it cuts down the time they have to spend in smog-ridden queues to pay a toll. They make an upfront investment in setting up the account and installing the transponder, and they get the daily benefit of a less frustrating commute. A green lane in a seaport would be authorized only for smart and secure containers whose integrity and location can be tracked. The benefits would come in three ways. First, the users of the green lane would be provided with assurances from U.S. authorities that these boxes would receive preferential treatment, which translates into a lower risk of inspection. If their shipment is targeted for inspection for any reason, it would be moved to the head of the line. Second, should the United States have to set a higher level of terrorist alert, the inspection rate of containers that had come through an overseas green lane would remain unchanged. Finally, should the U.S. government have to temporarily close down its ports following a terrorist attack, the first containers that would be allowed to move again once the ports were reopened would be those that originated from secure ports or terminals that have green lane privileges.

For a green lane to attract users there would also have to be a red lane. Boxes that arrive at an overseas port without any of the new safeguards would have to be subjected to increased inspections. At a higher security-alert level, these boxes would either be prohibited from being loaded or required first to be brought to an accredited security-sanitized facility near the port where, for a fee, these goods would have to be unloaded and reloaded into an approved box.

The green lane-red lane plan capitalizes on the same "time is money" market force that has been undermining traditional border controls for decades. By using delay as a stick and facilitation as a carrot, the transportation system can be adjusted to redistribute the economic rewards for good security practices, as opposed to sweeping them aside to drive down costs. Adopting smart and secure containers becomes the only way to stay competitive.

Undertaking such an ambitious approach to securing the trade and transportation system would have been a nonstarter before 9/11. I remember trying to make the case for harnessing supply-chain visibility to support border control to a crusty veteran customs supervisor on California's border with Mexico in January 2001. He politely heard me out and then said, "Commander, the only thing I will ever trust is the nose of a customs inspector." Underpinning his skepticism was a long-standing conviction by frontline agencies that the only sure path to security was more inspections on U.S. territory, where we could enforce our laws. Veteran agents and managers viewed as unworkable the concepts of engaging the private sector as a partner, pushing our borders out to police imports far from our shore, and harnessing technologies to monitor the status of goods heading our way.

September 11 has given these new ideas some impetus, but there still is not enough resources and urgency to move beyond an enhanced version of the "trust but don't verify" system that has survived the attacks on the World Trade Center and the Pentagon. The limited progress that has been made is due to the tenacious efforts of Robert Bonner, who became commissioner of the U.S. Customs Service just eight days after the 9/11 attacks. Immediately after being sworn in, Bonner declared that the primary mission of his agency would be combating terrorism. A former U.S. Attorney and U.S. District Court judge, Bonner served as the head of the Drug Enforcement Administration (DEA) when George H.W. Bush was in the White House. As the top official at DEA, he had gained a hands-on understanding of the relative ease with which smugglers could get contraband into the country.

Bonner had read a post–9/11 essay I had written, titled "The Unguarded Homeland," and asked to meet with me in early December 2001. He was very interested in my recommendations for transforming the way we police transportation networks and the nation's land and sea borders. We had an extensive discussion particularly on my proposal to exploit the untapped potential of the world's largest ports to advance trade security. A little over a month later, in a speech at the Center for Strategic and International Studies, Bonner announced what he called the Container Security Initiative (CSI). The largest container ports in the world would be approached to host U.S. customs inspectors so that boxes could be targeted for inspection before they were loaded on a ship bound for the United States, as opposed to after they arrived. He also extended an offer of reciprocity to any participating country. If they agreed to host our inspectors, we would agree to host theirs. Bonner argued that this approach offered the best hope of balancing the trade and security imperatives, adding, "As with any new proposal, implementation of this initiative will not be easy. But the size and scope of the task pale in comparison with what is at stake. And that is nothing less than the integrity of our global trading system upon which the world economy depends."

The Container Security Initiative is the companion piece to a program that Bonner announced in late November 2001, called the "Customs–Trade Partnership Against Terrorism" or C-TPAT. Under C-TPAT, the customs commissioner has tried to enlist the trade community as a counterterrorism ally. The havoc caused by the near closure of U.S. borders immediately after September 11 had gotten their attention. So Bonner took a page from President Bush's "you're either with us or against us" book. Companies that routinely imported goods into the U.S. were told that they needed to take a good look at the potential vulnerabilities within their supply chains and develop a plan to address them. Importers who chose to pursue business as usual were told they would find themselves in the cross-hairs of an increasingly no-nonsense customs service, and they could look forward to associated delays, audits, and stiff fines for infractions.

Commissioner Bonner also has changed the long-standing practice of submitting cargo manifests at the port of entry instead of at the port of departure. In addition, he has disallowed cargo declarations that use such vague descriptions as "Freight All Kinds" (FAK) or "General Merchandise." As of December 2, 2002, ocean carriers are required to electronically submit a cargo declaration twenty-four hours before cargo destined for the U.S. is loaded aboard the vessel at a foreign port.

This twenty-four-hour vessel-manifest rule is important, because without it, there is no credible way to run a risk-based targeting program. From the standpoint of intercepting a shipment, if the U.S. government has intelligence that there may be a danger, it is important to be able to act on that information before a container arrives in the U.S. If word of a container's arrival comes after it is already here, it may be too late if a terrorist has armed it to go off in the port. Also, having U.S. customs inspectors assigned overseas as a part of the Container Security Initiative only makes sense if they have cargo information before the ships depart from their port. Otherwise the only thing they can do is randomly select containers to inspect, which boils down to a largely hopeless needle-in-the-haystack exercise.

The trade community initially expressed considerable consternation about the new twenty-four-hour rule, complaining it would raise their costs and create delays. However, once the enforcement deadline of May 3, 2003, arrived the program was up and running with few hitches. In addition, by the summer of 2003, all of the twenty largest container ports agreed to participate in the Container Security Initiative, and there are nearly an equal number of smaller ports that have expressed an interest in joining in a second phase of the program announced in May 2003. Further, by the end of 2003, 4,600 importers, ocean carriers, and freight forwarders had submitted applications to join C-TPAT.

The speed with which these new initiatives have been embraced is easy to explain. It stems from the fear of both importers and foreign port authorities that U.S. inspectors will subject shipments from nonparticipating companies and ports to greater scrutiny with the associated delays. Unfortunately, these fears are largely unfounded, because the Bureau of Customs and Border Protection lacks the manpower and resources to adequately staff the Container Security Initiative, to review the applications of companies who wish to participate in C-TPAT, and to move away from error-prone cargo manifests that remain the cornerstone of its targeting system. The carrot of facilitation that comes from participating in these programs is not matched by a credible stick. And none of these programs address the core cargo security imperative of confirming that the goods loaded into a container from the start are indeed legitimate and that the container has not been intercepted and compromised once it is moving within the transportation system.

Two initiatives that I helped to launch after 9/11 have led to real-world tests in how new processes and technologies might be adopted to secure and track containers from their point of origin. The first is the publicly funded Operation Safe Commerce which began with a relatively simple proof-of-concept test in which a global-positioning-system device and intrusion sensors were attached to a shipment of automotive light bulbs. The container originated in a manufacturing plant in the East European country of Slovakia, traveled by truck to the port of Hamburg, Germany, then by ship to Montreal, and then by truck to a factory in Hillsborough, New Hampshire. This small demonstration project took place in May 2002 and helped to spawn what is now a $58-million program managed by the Transportation Security Administration and operating out of the three largest maritime trade centers in the United States. The port authorities in the New York, Los Angeles, and Seattle areas are undertaking a variety of tests designed to analyze the feasibility of routinely monitoring the supply-chain integrity of U.S.-bound cargo container shipments that originate from the interior of Asia and Europe. Unfortunately, the White House's budget proposal for 2005 provides no funding to continue this program.

The second project is a privately funded effort known as the Smart and Secure Trade Lanes Initiative and involves a consortium of sixty-five companies. The organizers of this project have collaborated with a number of international organizations in running a series of pilot projects, including the World Customs Organization, the International Standards Organization, the Asia Pacific Economic Community, and the European Union. The participants believe there is a business case, as well as a security justification to be made for developing a global means for tracking the location and status of the contents of a container. The preliminary results are quite encouraging. During the first year, they demonstrated they could keep track of both the location and all the associated documentation for over eight hundred containers. Based on preliminary economic analysis, they found that a shipper moving $70,000 worth of goods saved, on average, $400, due to reduced operating costs and improvements in inventory management.

While the U.S. government's investment in all of these initiatives continues to be inexplicably modest, there remains a fertile basis for attracting and sustaining cooperation with the private sector and the international community. For private companies, their current business practices simply cannot survive, should the U.S. government respond to the next terrorist attack on American soil by throwing a global kill switch. Even if the transportation system must be shut down in the immediate aftermath of an attack, companies have a vested interest in having it turned back on as quickly and efficiently as possible. That will require having a credible security regime in place with which to convince a traumatized American public that it is reasonably safe to move cargo.

Other countries should be supportive of improving trade and transportation security, even if they do not feel they are likely to be targets of terrorists. Every nation has something it defines as contraband. Containers are used to smuggle weapons, drugs, cigarettes, migrants, child pornography, and every other kind of prohibited item, and criminals are always looking to diversify their markets. Also, all civilized nations will want to avoid tragedies like the one in the summer of 2000, when fifty-eight Asian migrants, trying to make their way into the United Kingdom, were found suffocated in a container.

Many public-health strategies aimed at managing the spread

of disease require the identification and isolation of livestock as well as agricultural products that could place the food supplies and the general population at risk. Safety and environmental threats connected with unsafe shipping and trucking also mandate that the transportation sector be monitored. Particularly in the developing world, many states continue to rely on the collection of import duties as an important source of government revenue. Poorer countries are routinely cheated of those duties by unscrupulous importers who mislabel, undervalue, and underquantify what they are importing. Since it is hard to figure out what is in the box, small countries like Jamaica estimate they are being defrauded of as much as 80 percent of the duties they should be collecting.

Developing the means to secure the transportation system from the new shadow warfare is one of the many daunting challenges facing the United States. The approach we need to take in order to manage it, and the stakes that go with getting it right, offer an instructive lesson in how much remains to be done to protect our way of life and the critical foundations of U.S. power. The magnitude and the complexity of the task highlight why we should show more skepticism about official assurances that modest security measures are yielding dramatic results.

We face a big challenge in providing adequate security to vital global networks. We must join with the best thinkers in the private sector and other countries in providing "big" solutions. If we don't, we may periodically discover that the answer to the question, "What's in the box?" is more death and disruption visited on American soil.

6
Protect and Respond

We are sitting on a time bomb. U.S. intelligence officials had initially believed that the planning process for the attacks on New York and Washington was only two years long. From interrogations of al Qaeda operatives who have been captured since 9/11, we now know that Osama bin Laden and his cohorts spent five years formulating the plan. In light of this revised time frame, the need for America to ward off complacency is more compelling than ever.

The threat is ongoing, and we must be doing more right now to limit the consequences of future acts of terror. It will take time to implement well-conceived, layered security measures that protect the critical foundations of our society. We have to make judgments about our most immediate vulnerabilities, identify what stop-gap protective measures we can implement in a hurry, and develop and exercise plans to guide our response when the next attacks materialize.

We should deal with our homeland security task as though

we were tackling a thousand-piece jigsaw puzzle. When you first dump out the pieces, it is easy to be intimidated. But you start by turning all the pieces face up, sorting them out by common features, and then beginning the slow process of piecing them together. In time, it becomes easier to decipher patterns; the picture takes form, and the process of assembling gets easier.

Turning over the pieces is where we need to start. Priority should be given to those areas in which an attack could cause great loss of life or profound societal disruption. If we have credible intelligence that terrorists appear intent on hitting specific targets, we should certainly mobilize protective measures for those targets right away. But we should recognize that our intelligence is likely to be spotty for some time to come. This means operating on the assumption that we will not get advance warning and that we can not have a specific threat-based approach to deciding what needs to be protected. Instead, we need to invest energy and resources to better protect assets that make for likely targets. At the same time, we should also have contingencies in place to manage the aftermath, should our defensive efforts still fail to deter an attack.

Container security is only one important area that should be commanding our attention. Another sector that rates top security billing involves a basic element of life: our food supply. The systems that provide food to our country's large population are diverse, complex, and fragmented. When Americans sit down to consume a variety of meats, seafood, fruits, vegetables, and processed foods each day, the dinning room is the final stop on a vast global chain of production and distribution. The raw materials start out on fishing vessels or farms around the globe and move on to processing plants or distribution chains. From there, they are transported to the shelves of wholesale or retail outlets, and then are purchased by restaurants or individual consumers for their meals.

All along the way, there are plenty of opportunities for a terrorist to introduce lethal contaminants into the food supply chain. They could target livestock where they are raised or crops where they grow. They could introduce a contaminant at a food processing facility, a livestock feedlot, slaughterhouse, or meatpacking facility, or into food products transported to market. Finally, they can target market-ready foods at a store, as occurred in January 2003 when ninety-two people became ill when a Michigan supermarket employee intentionally contaminated ground beef with substantial amounts of nicotine. Restaurants are also vulnerable, as was demonstrated a decade ago after an obscure cult spread salmonella bacteria in restaurant salad bars in a small town in Oregon in order to sicken voters in advance of a local election. While the danger is real and the target is relatively soft, most of the people who work in the food industry still feel they lack the information to identify the most serious threats. Nor has this sector been mobilized to think about and prepare to respond to a deliberate act of sabotage or bio-terror.

What makes the lack of preparedness within this sector so worrisome is not just that an attack on the food supply system might poison thousands of Americans. Like a dirty bomb in a shipping container, an act of agroterrorism could also have enormous economic consequences. Food production makes up nearly 10 percent of the U.S. Gross Domestic Product. It generates nearly one trillion dollars in cash receipts and employs one in eight American workers. It is also an important component in the U.S. balance of trade, since our agricultural exports are double the total of other U.S. industries. One California study estimated that an outbreak of foot-and-mouth disease could cost the state one billion dollars in lost trade. But this kind of epidemic would not be limited to California. A recent simulation informs us that by the time the Agriculture Department's foreign-disease laboratory on Plum Island, New York, would have confirmed the first case of foot-and-mouth, it would likely have spread to twenty-eight states, at which point most of America's $90-billion livestock industry would have to be wiped out.

On December 23, 2003, the U.S. beef industry learned firsthand what the discovery of a single cow infected with mad cow disease can do to their industry. News that an infected cow was found in the state of Washington led to an immediate ban by thirty nations on all U.S. beef exports. In so doing, they closed their markets to an industry that generated $3.2 billion in annual export revenues. This case was discovered on a small dairy farm. If a contagious disease such as foot-and-mouth broke out in the cattle herds around Amarillo, Texas, up to 1.5 million head of cattle located within a hundred-mile radius would have to be slaughtered. Destroying that many animals is obviously a daunting and extremely costly task. By way of illustration, it would take a ditch more than seven hundred miles long to bury that many cows lying side by side.

As the mad cow incident highlighted, there is no real-time communications network for the U.S. Department of Agriculture (USDA) to manage a disease outbreak with state authorities or with U.S. trade partners. A major outbreak could easily involve dozens of agencies with no one clearly in charge. Confusion over who has an obligation to file reports, who has jurisdiction, and just what the appropriate response should be to an agroterrorist attack promises to seriously compromise public confidence. This problem will be compounded by the fact that there is such limited capacity to conduct laboratory tests for exotic contaminants. Without testing, it would be impossible to quickly reassure an anxious public that their food is safe, once a highly publicized incident takes place.

Introducing adequate security to protect our food supplies should be a top national priority. There are a number of actions that should be pursued to reduce the risk. First, key relationships need to be strengthened among the constituents throughout the food industries. There have been long-standing tensions among farmers, feed yards, packing companies, food processing plants, and retailers as they all jockey to maintain a piece of the industry's thin profit margins. Within each sector, small players often see themselves as underdogs who are being rolled over by big conglomerates—a perception that is not far from reality, as the industry becomes increasingly concentrated. What is missing is the kind of regular forum where operators meet with security experts as they now do at Logan Airport in Boston.

There is some sign of progress that the industry as a whole is trying to develop the capacity to speak with one voice. One promising new organization called Veriprime was launched in September 2003. The group brings together beef producers, processors, and restaurant chains that have agreed to share best practices among themselves to advance food safety. The federal government should actively encourage and support the formation of this kind of group.

Second, federal, state, and local public health agencies, law enforcement agencies, farm bureaus, and local and state public officials have to create regional committees to develop protocols for how they would respond to a terrorist attack on the food supplies. This group must develop an effective public information plan in the event of a serious viral or bacterial outbreak. Information is essential in managing public anxiety. Providing timely reports of what is happening and authoritative guidance on what the public should do is a must, and the message should not be muddled or contradicted by officials and commentators operating by the seat of their pants.

We should also create a food-health-personnel reserve system in which medical professionals, veterinarians, and plant pathologists can be mobilized and sent to support these committees in the effected areas. A model for this can be found in North Carolina, where the state's small emergency-preparedness staff of thirty-five people has created county-level animal response teams of volunteers. These teams include animal control officers, university researchers, the sheriff's departments, veterinarians, forest rangers, animal producers, and private citizens.

Third on the list of immediate actions is to conduct regional exercises to test the ability of sectors to detect and respond to an attack. Field exercises always identify critical holes in contingency plans and strengthen relationships among the participants. They also make the issue much less abstract for the key decision makers who must be counted on to play a helpful role in sustaining political support for resources. Instead of one national exercise every two years, we should invest in at least a dozen exercises around the country right away, and then six per year thereafter.

The fourth objective is to harness technologies to monitor the food supply in near real-time. These devices should be routinely operating at feed centers and food processing plants before food is sent to consumers. Radio-frequency identification devices should be used to monitor agriculture shipments so that it is possible to quickly identify their point of origin. Efforts by the USDA to use electronic tagging and sophisticated database systems to track cattle from farm to feedlot to factory are an excellent example of how off-the-shelf technologies can be harnessed to improve transparency in the food supply. If a diseased cow is identified, these tags will allow authorities to quickly verify the other animals that might have come into contact with it so the contaminated livestock can be quarantined and vaccinated within forty-eight hours. (In the case of the Washington State mad-cow, it took four days to determine that the diseased cow had come from Canada and fifteen days to confirm that fact through DNA tests. A month and a half after the discovery, the USDA had failed to find two-thirds of the animals that were at risk of infection when they closed their investigation.) When authorities have to order a culling of livestock, they should be able to establish reasonable boundaries on the destruction of herds. All these technologies are partly about identifying and containing a disease outbreak, and partly about having the means to reassure an anxious public that the government has a credible basis for distinguishing what food is still safe to consume. Such matters should be pursued with haste.

Finally, there has to be investment in laboratory capabilities to support local farm bureaus and public health authorities. At the federal level, Congress should create a dedicated agricultural disease counterpart to the Centers for Disease Control and Prevention (CDC) to guide research and prevention efforts, and to maintain international contacts to monitor disease outbreaks and new bioterror developments overseas. Testing should be conducted routinely and on a large-enough scale to provide a form of early warning. A uniform system for electronically reporting disease and infestation should be nationally integrated and accessible to frontline medical and veterinary professionals. Also, laboratories should have the capacity to handle the surge in demand for testing associated with a major infestation or outbreak.

Along with the security of our food supplies, it is crucial that we dramatically improve security of the chemical industry. Our enemies do not need to smuggle chemical weapons across our borders. Just as the 9/11 attackers succeeded in converting domestic commercial aircraft into missiles, chemical facilities and the thousands of tons of chemicals that move each day around the U.S. on trucks, trains, and barges could be targeted by terrorists to devastating effect.

All told, there are about 15,000 chemical plants, refineries, and other sites in the U.S. that store large quantities of hazardous materials on their property. Many are in industrial parks, some distance away from residential populations. Yet there are important exceptions to the rule. For instance, public water utilities are among the top three consumers of chemicals, including chlorine, which is one of the most hazardous, and these chemicals are typically stored on sites near treatment facilities. Detonating a tank of chlorine gas can potentially lead to the death or injury of tens of thousands of people who are downwind. Historically, there has been little in the way of physical security to monitor or guard these facilities. In addition to the lives lost from chemical releases, targeting waste-water treatment plants could also lead to system-wide shutdowns.

According to the Environmental Protection Agency (EPA),

there are 823 sites where the death or injury toll from a catastrophic disaster at a chemical plant could reach from 100,000 to more than one million people. Admittedly, these are worst-case scenarios. Still, given all the hand-wringing done over nuclear power plants, the level of long-standing indifference when it comes to the chemical industry is astounding.

One of the main reasons that the chemical industry is for the most part out of sight and out of mind is that on the whole, it has an admirable safety record. Each day this industry handles the most lethal substances known to humanity and converts them into benign uses that are essential to the functioning of our economy. Because of its track record, the industry has been allowed to largely self-regulate, with some oversight by the EPA and the Department of Transportation.

There are no federal laws that establish minimum security standards at chemical facilities. After 9/11, Senator Jon Corzine of New Jersey drafted legislation that would require chemical companies to identify the vulnerabilities in their operations and prepare security plans to address them. These plans would have to be submitted to the EPA or Department of Homeland Security for approval. The bill also called for the EPA and the new department to issue new rules that compel some plants to use safer procedures if they operate in high-risk areas. The chemical industry rallied nearly thirty trade associations from manufacturing and agricultural groups to oppose these new requirements. As a result, a more industry-friendly bill sponsored by Senator James Inhofe was introduced that requires owners of selected chemical storage facilities to conduct vulnerability assessments, while allowing the Department of Homeland Security to only access them upon request. Many in the chemical industry are likely to be satisfied with this compromise since they know, in the absence of a regulatory mandate, that the department will not be staffed to carry out these reviews.

So while our chemical industry may be safe, no one can accuse it of being secure. Basic measures such as posting warning signs, fencing, and maintaining access control and twenty-four-hour surveillance are only required for 21 percent of the 15,000 sites that store large quantities of hazardous materials. Chemical railcars routinely sit parked for extended periods near residential areas or are shipped through the heart of urban centers. For instance, each week shipments, including deadly chloride, are carried on slow-moving trains that pass within a few hundred yards of Capitol Hill in Washington, D.C. Chemical barges that move up and down our inland waterways are unmonitored. If terrorists were to target them as they sat in a canal lock system or passed under a bridge, they could cause long-term disruption to commercial traffic on these vital river highways.

In short, the chemical industry deserves urgent attention because the stakes are high, the opportunities for terrorists are rich, and no credible oversight process exists. It is the very ubiquity of the U.S. chemical industry that gives it potential to be a serious source of national alarm. The morning after the first terrorist strike on this sector, Americans will look around their neighborhoods and suddenly discover that potentially lethal chemicals are everywhere, and be aghast to learn that the U.S. government has still not developed a plan to secure them. The subsequent political pressure to shut down the industry until some minimal new safeguards can be put in place—as we did with commercial aviation following the 9/11 attacks—will be overwhelming.

Among the immediate steps that must be taken is the requirement that vulnerability studies and security plans be completed and submitted to the federal government for review. The public is simply not going to accept self-validation as adequate after the first successful attack provides prima facie evidence that a security plan was not up to snuff. They will want some form of audit from a source that is committed to the public interest as opposed to a purely private one. Further, these security plans must be actively tested by conducting not only inspections but simulated attacks. Given the mortal risk to so many people, Americans have a right to expect robust security in this industry, not a pro forma exercise of checking boxes on a generic checklist.

Additionally, all trains, trucks, or vessels carrying hazardous chemicals should be equipped with GPS devices and tracked in a secure network as they move through the national transportation system. The routes should be reviewed to minimize the time they spend in populated areas. Extremely hazardous rail shipments, such as chlorine gas, should be rerouted around cities. Comprehensive exercises should be staged throughout the U.S. to test the capacity for various regions to respond, should the worst happen. Finally, Congress should reconsider Senator Corzine's proposed provisions to end the use of some especially deadly chemicals at plants near high population areas.

In addition to the threat arising from the chemical industry, there is good cause to be concerned about the danger posed by dirty bombs in which radiological materials are released by detonating a conventional explosive device. A dirty bomb is far more likely to be a weapon of choice by our enemies than a nuclear warhead mounted on a missile. It is also far less costly to take measures that control this risk. A study prepared by Michael

Levi and Robert Nelson of the Federation of American Scientists in 2002 outlined three categories of recommendations that deserve our immediate attention.

As a starting point, the U.S. government needs to work with companies, medical centers, and universities to reduce the opportunity for terrorists to get their hands on radioactive materials. Plutonium, americium, cobalt-60, and cesium-137 can be found in thousands of devices used for everything, from detecting smoke to disinfecting food and treating cancer patients. Significant amounts of these materials can be found in research laboratories on college campuses, hospitals, food irradiation plants, and oil drilling facilities. Securing these dangerous materials requires improving the capacity to monitor them.

Federal agencies, including the Departments of Energy, Health and Human Services, and the Nuclear Regulatory Commission, need to be funded at adequate levels to carry out inspections of the security surrounding these materials, including physical protection measures and oversight of recordkeeping. When the devices that contain these materials are no longer in service, the Department of Energy is responsible for collecting and disposing of them. But the budget for this disposal program is going in the wrong direction, having endured significant cuts since 2001. This trend needs to be reversed immediately so these materials can be taken out of circulation and destroyed as soon as possible.

There also needs to be substantial effort to deploy detection devices capable of identifying dangerous amounts of radiation across the transportation system. These include radiation detection sensors built within containers; radiation detection portals operating in domestic and overseas ports as well as commercial U.S. land border crossings; and sensors located at places such as rail yards, tunnels, along highways, and near national landmarks. Detectors also should be installed and networked within the urban areas most likely to be targeted by terrorists.

Another imperative for combating terrorist threats is preparing first responders and hospital personnel to cope with incidents involving chemical or biological weapons. Firemen, police officers, and medical personnel need to know the facts surrounding radiation exposure and what appropriate steps should be taken to protect and decontaminate themselves and others. Getting everyone to sing from the same sheet of music is no small undertaking, given both the number of local, state, and federal organizations involved and the fact that there are 2.7 million nurses and over a million firefighters and police officers nationwide. Internet-based training tools should be used to help ensure consistency and quality control over the training curriculum.

It is not just radiological materials that need greater controls. Even deadlier materials could be used in acts of bioterror, yet there is no federal program to provide ongoing oversight of how lethal pathogens are handled. Many university research labs around the country hold highly contagious specimens, but too often the level of security they practice could only be charitably described as lax. In 2002, federal inspectors found seven vials of the pathogens that cause bubonic plague and pneumonic plague in an unlocked refrigerator. The vials were notionally in the custody of a university lecturer who had not done an inventory of the freezer since 1994. When pneumonic plague becomes airborne it is almost 100-percent fatal. Its victims usually die within forty-eight hours.

What is needed right away are federally mandated standards that require security plans for labs that hold these dangerous substances. There should be adequate locks and surveillance cameras installed. Access should be limited to lab workers whose backgrounds have been checked. Universities and companies also need to maintain a centralized database of the biological materials in their custody, and there should be mandatory reporting when pathogens are discovered missing. All of this needs to be monitored by government inspectors, who should be routinely deployed to check on the state of security.

Getting our own security house in order still won't help with the challenge of preventing bioweapons from finding their way into the U.S. from overseas or getting cooked up in a kitchen laboratory. One frightening possibility is that new advances in biotechnology might lead to the creation of highly virulent biological agents. The potential dawning of this new age was suggested by a team of biologists that recently created a successful polio virus in vitro. Since the technology involved is used for legitimate research, it will be very difficult for intelligence officials to detect a nefarious effort to create a deadly biological agent.

We should assume that we will not be able to detect and intercept all efforts to strike Americans with biological weapons. Given that these agents would be difficult to control under the best of circumstances—and current conditions are a long way from ideal—we should be making an urgent effort to bolster our local and state public-health capabilities to detect outbreaks of disease early and to treat the victims.

At the federal level, the Bush administration is putting in place a system called Bio-Watch, which involves installing and monitoring a nationwide set of air sensors to check for the presence of smallpox, anthrax, and other pathogens. The federal government is also in the early stages of developing a public health surveillance system built on monitoring the health databases of eight major cities for signs of disease outbreak. Finally, the Strategic National Stockpile has been boosted so that adequate medications can be rushed to the airports of affected cities within twelve hours.

These are positive steps, but as with so many of our homeland security efforts, they come with too few resources to address the need. Surveillance systems should be up and running in all our major metropolitan areas. Americans are always on the move. If a biological weapon is released in an urban area that is not being monitored, a contagious disease could spread into multiple states before the first alarm is sounded.

It is also worrisome that stepped-up federal efforts are generally not being matched at the state and local levels. Public health agencies have not been a high priority in recent decades, and lean state and local budgets since 2001 have only made matters worse. Most health agencies are barely staffed to run during a normal 9-to-5 workday. Further, states' public health reporting systems are largely antiquated, slow, and outmoded. It can routinely take up to three weeks for a public health department to get a disease incident report to reach the Centers for Disease Control. The CDC has just begun undertaking the process of developing a common language and diagnostic coding system to support the operation of a national database.

Even if a bio-attack is detected early and the federal stockpile of medications is shipped to the airport of the targeted city right away, our troubles are still not over. According to a study of emergency responses to a hypothetical anthrax attack, completed in 2003 by Lawrence Wein and Edward Kaplan, the release of just two pounds of weapons-grade anthrax dropped on an American city could result in more than 100,000 deaths. The reason for this huge death toll is that the distribution of antibiotics at "street level" would be too slow to treat victims early enough, and once people develop symptoms, they would overwhelm the capabilities of medical facilities. These findings were confirmed in a classified exercise run by the federal government in the fall of 2003.

A big part of the problem is that the need for developing hospital capacity to deal with mass casualty events runs head-on into the conflicting imperative to contain rising health-care costs. In recent years, medical center administrators have worked hard to drive down overhead costs by streamlining their operations. One consequence is that they have little in the way of surge capacity, and most do not maintain on hand the specialized medical equipment needed to cope with nonroutine events, such as a bioterrorist incident. An August 2003 report on hospital preparedness, put out by the U.S. General Accounting Office, found that most urban hospitals had a shortage of equipment, medical stockpiles, and quarantine and isolation facilities for even a small-scale response to a contagious disease outbreak. Half the medical facilities in the U.S. do not have enough ventilators to treat multiple patients who have the kind of severe respiratory problems brought on by anthrax or botulism. The American Hospital Association estimates that it would take an investment of $8 billion to bring all metropolitan hospitals up to a point where they could provide acute care in the event of a nuclear, biological, or chemical attack.

As a nation, we should make the investment to ensure that our public health agencies, emergency rooms, and hospitals are prepared to deal with the reality that we will probably have terrorist attacks in the future involving weapons of mass destruction. But it is not only our medical sector that is unprepared to cope with mass-casualty attacks on U.S. soil. Our fire and police departments would almost certainly be overwhelmed as well. As Martin O'Malley, the mayor of Baltimore, observed in April, 2002:

> Today, we are fighting a different kind of war—on two fronts. One front is Afghanistan, where we have the best technology, the best equipment, the best intelligence being sent right to the front, and no expense is spared. But for the first time in nearly 200 years, the second front is right here at home. And to date, it's where we've seen the greatest loss of life. Yet we have insufficient equipment, too little training, and a lack of intelligence sharing with federal authorities.

There are many challenges for our first responders. Topping the list is communications. In virtually every city and county in the U.S., there is no interoperable communications system to facilitate police, fire departments, and county, state, regional, and federal response personnel communicating with one another during a major emergency. Radio frequencies are not available to support the communication demands that will be placed on them, and most cities have no redundant systems to use as backups. Portable radios will not work in highrise buildings unless the buildings are equipped with repeater systems. Most U.S. cities have separate command-and-control functions for their police and fire departments, and almost no coordination exists between the two organizations. Furthermore, with few exceptions, first-responder commanders do not have access to secure radios, telephones, or video conferencing capabilities that can support communications with county, state, and federal emergency preparedness officials or National Guard leaders.

There is also a shortage of protective gear and portable detection equipment. In the event of a chemical attack, a window of a few minutes to two hours exists to respond to the incident before death rates skyrocket. Yet protective gear is often available only to a few specialized incident response teams. Most communities report that within six hours, they will run out of even the most basic emergency resources, such as life-saving equipment, personal protective suits, oxygen, and respirators. Police officers who would be called on to secure an incident scene would be particularly exposed since they are not outfitted with protective gear. Federal agency response teams can help, but they will almost always be too late, since plans call for them to arrive no sooner than twelve hours after the attack. Portable detection equipment for deadly chemical, biological, and radiological materials is in short supply and is notoriously unreliable in urban environments. In many cases, the sensors that have been issued to local first responders have arrived without adequate personnel guidance on how to use and maintain the equipment, or even what to do, should the detection equipment register an alarm.

Then one must consider training. Major field exercises are important tools to test the adequacy of contingency plans, equipment, command-and-control procedures, and training. In all but America's largest cities, there is a paucity of resources and expertise to conduct these large-scale exercises. Important specialized training is also in woefully short supply. For example, the Center for Domestic Preparedness in Anniston, Alabama, is the only facility in the nation in which first responders can train with and gain firsthand knowledge of chemical agents. At peak capacity, it can train only 10,000 first responders per year.

Part of the reason why the first-responder situation is so grim is that there are no established preparedness standards to identify baseline requirements. Without agreed-upon standards, localities oscillate between inertia and scattershot spending. As the July 2003 Council on Foreign Relations report has recommended, the federal government needs to reach out with a sense of urgency and work with state and local agencies, as well as emergency responder associations, to establish guidelines for emergency preparedness. While there cannot be a one-size-fits-all approach, there should at least be agreement on minimal communications, equipment, and training capabilities that frontline responders should possess to safely work with one another while coping with a massive attack involving deadly agents.

There is a category of threats not involving weapons of mass destruction that could still result in serious disruption to our daily lives. These include attacks on the cyber, financial, and energy backbones of our society. In each of these areas, the greatest threat is not so much the attack itself, but the economic harm that arises from a loss of public confidence in these systems. When an attack brings the depth of our dependency on infrastructure we otherwise take for granted to the forefront, people and companies start to hedge their bets.

In the cyberworld, people will be more cautious about communicating over the Internet and more hesitant to engage in commercial activities like online shopping. The financial industry is particularly dependent on maintaining public trust. If banks fail to protect their customers' assets or privacy, people will start thinking about putting their money under their mattresses. Regular blackouts mean that companies need to incur the expense of buying and maintaining emergency generators.

To a certain extent, tempering our reliance on the critical foundations of our society is a prudent response to the dangers that confront us. However, a wholesale loss of public faith in the dependability of these networks would be devastating. Accordingly, in each of these sectors there are things we have to do right away to reduce that risk.

We need to stop dancing around the imperative of establishing security standards. There have to be standards and there has to be an adequate capacity to vigorously enforce them. Relying on best practices and industry self-policing was acceptable for meeting our pre–9/11 regulatory needs, but they are simply inadequate in the post–9/11 security world.

How we arrive at those standards is a legitimate topic for discussion. What will not work is government agencies cooking up new mandates behind closed doors. The people on the government payroll who actually understand the full complexity of these sectors are few. As a result, the kinds of prescriptions that agencies gin up on their own will likely have the downside of being costly and not providing any added measure of security. The way around this conundrum is to put industry in the lead for devising the security requirements, and then have the government validate them and oversee their enforcement.

There should also be ongoing incentives for exceeding minimal requirements. The idea is to create an incentive structure similar to the one that operates in the consumer credit market. In general, people who want to borrow money can do so as long as they don't have a pattern of defaulting on their loans. But borrowers who are careful not to stretch themselves too thin among creditors and who consistently pay their bills on time get better rates than those who don't. This kind of approach is gathering steam in the area of environmental protection. Those companies who demonstrate that they are willing to go the extra mile in embracing sound environmental practices get treated better by government oversight agencies than those who take a minimalist approach. Good companies are promised fewer inspections or reduced reporting requirements, which lowers their regulatory compliance costs. Such an approach is taken by the Coast Guard to encourage ships to exceed the international safety standards designed to help prevent oil spills.

Offering carrots for those who are committed to doing more on the security front must be matched by sticks for those who try to do less. Oversight agencies have to possess the staff, training, and resources to ensure compliance. If they don't, the program will lack credibility.

Moreover, managing the threat to our critical infrastructure has to include a well-practiced plan for responding when something goes wrong. If a network needs to be brought down, the mechanics of doing it should be worked out in advance. Equally important is the protocol for bringing the unaffected segments of the network back on. This is a case in which industry and government must plan for the worst. Unfortunately this has not been done in the transportation and energy sectors. For instance, if the U.S. government decided to close our seaports following a terrorist attack, agencies still have no plan for how they would get traffic moving through ports once they elected to open them up again. Given the likelihood that such an attack will take place, a make-it-up-as-you-go approach to restarting the global transportation system is not likely to inspire much in the way of public confidence.

This list of urgent homeland security needs and recommendations is by no means exhaustive. Nonetheless, the discussion highlights the scale of the challenge we are facing. There is an understandable reluctance by elected officials to honestly discuss the enormity of the homeland security enterprise. Most of the rewards in Washington come from quick wins and not from tackling long, often thankless, tasks. Many public officials I have met have been defensive when I suggested that a greater sense of urgency needs to be brought to bear to tackle the threats that confront us. For some, this comes from a fear of their political exposure if they were to acknowledge the limits of what they have been able to achieve. For others, their defensiveness grows out of a legitimate sense of pride about what they have accomplished in the short span of time since 9/11. But whether it is fear or pride, I am convinced they are setting themselves up for a big fall when the next attack puts what they have been doing—or not doing—under a magnifying glass.

The challenges we face did not materialize overnight. They are a consequence of decades of work in constructing the foundations for our society, without anyone paying much heed to the imperative of security. Reducing our vulnerabilities has been further complicated by the fact that these foundations are increasingly connected to global networks. The result is both to heighten our exposure and to increase the number of interested parties who must be involved in formulating a response. It is unreasonable for anyone to expect that the current occupants of the seats of power in Washington could turn all this around in a year or two. This is a historic undertaking.

Some may think that the undertaking is simply too expensive or too hard. Others may believe that it can be sidestepped altogether if we are assertive enough in going after those who intend to harm us. But the dragnet we cast overseas can only be one part of a layered approach to our security. We have to ensure that the critical systems that support our way of life are harder targets to hit. The effort will bring residual benefits in terms of safety and the combating of crime, and in some instances, it will generate efficiencies as well. But more important, by embracing the homeland security agenda, we can deter many of our enemies from trying to attack us in the first place. We can do this by denying them the kind of odds they now enjoy in creating a big disruptive bang for their buck.

We cannot dodge the homeland security enterprise. Nor should we be overawed by it. Many of the challenges are issues of policy and coordination. It will take determined executive leadership to keep us on course. Most of all, it will take mobilizing the nation by providing a framework for the sustained engagement of the American people in safeguarding the nation.

7

Mobilizing the Home Front

One of my more painful moments as a patrol boat commander was when I chased two high-speed vessels around Virginia's lower Chesapeake Bay, having mistaken them for drug smugglers. They turned out to be undercover boats operated by the Customs Service and the Drug Enforcement Administration. Once I intercepted them, I found out that they had received the same intelligence tip I had but had failed to alert the Coast Guard of their plan to try and make the drug bust. If the smuggler had actually been heading toward the reported rendezvous point, he undoubtedly changed his plans upon seeing my white Coast Guard cutter with its red racing stripe in hot pursuit of a cabin cruiser and a speedboat that were manned by other law enforcement agents.

Several months after 9/11, I related this story to a group of academics who were debating whether the likely pain of creating a new Department of Homeland Security (DHS) was worth the potential gain. My message: it would be difficult to imagine how any reorganization effort could make matters worse.

In testimony before Congress one month after the attacks on New York and Washington, I provided another illustration of how badly broken things were on the front lines. I told the Senate Governmental Affairs Committee on October 12, 2001, that, if the 9/11 attacks had come by sea instead of air, only dumb luck would have prevented a ship with a shady past, carrying suspect cargo, and manned by a questionable crew, from entering a U.S. port. This was because it was impossible for multiple red flags to be viewed simultaneously. The Coast Guard might have known something about the ship. Customs may have had the cargo-manifest information if it was submitted in advance (pre-arrival notification was optional then). Immigration officers may have known something about the crew—depending on the accuracy of the crew list faxed to the Coast Guard by a shipping agent and the speed and accuracy with which the names could be manually loaded into a database. Moreover, none of the frontline inspectors in these agencies had access to national security intelligence from the FBI or the CIA. And all of these inspectors confronted more people, cargo, and ships that sparked their interest and concern than they had the wherewithal to intercept and inspect.

These glaring shortcomings were well known inside the U.S. government before September 11, but senior policy makers had been reluctant to take on the structural problems that gave rise to them. Instead, they took solace from the fact that terrorists were occasionally intercepted at the border. One of the celebrated success stories was the arrest of Ahmed Ressam, the millennium terrorist. On December 14, 1999, customs agents assigned to a ferry terminal in Port Angeles, Washington, arrested Ressam, an al Qaeda operative, when he tried to enter the United States from Canada with a trunkload of bombmaking materials. His target was Los Angeles International Airport.

The circumstances surrounding the Ressam arrest should have been cause for alarm, not self-congratulation. His capture was the border-control equivalent of winning the lottery. Among the many lucky breaks was the fact that Ressam had contracted malaria when he was learning the ABCs of terrorism in one of Osama bin Laden's Afghan camps. After a ferry trip from Victoria, British Columbia, he arrived feverish and groggy. So this well-trained operative, who had successfully crossed international borders dozens of times, and who had smuggled high explosives past U.S. immigration officials in Los Angeles, and who for years had lived off the proceeds of his shoplifting from stores and pick-pocketing tourists, appeared nervous to Diana Dean, the U.S. Customs Inspector.

The second bit of good luck was that Ressam's rented Chrysler was the last one to leave the ferry, so when Dean leaned into the window to ask where he was from and where he was headed, she wasn't under any pressure to wave him on. Other agents who had processed all their vehicles gathered from nearby. When Ressam handed over a Costco discounter card in response to a request for some identification, he gave another clue that he might rate a closer look. When he was asked to open his trunk, Ressam bolted from his car and ran off the ferry. That move alerted Inspector Dean's colleagues, who gave chase and tackled him minutes later.

In Ressam's trunk was a stash of 112 pounds of Urea fertilizer, two jars of EDGN—a gooey brown liquid explosive closely related to nitroglycerin—and the explosives RDX and HMDT in white powder form. These advanced explosives could have been used to make as many as four bombs the size of the one that destroyed the Murrah Federal Building in Oklahoma City. But the customs agents initially thought they had made a drug seizure. Contemporary newspaper accounts detail how Ressam kept ducking to the floor of the squad car whenever agents picked up the material, which, of course, turned out not to be heroin or cocaine.

When the drug tests came up negative, someone suggested that they should be tested as explosive materials. Unfortunately, Port Angeles had no such testing kits on hand. There ensued a debate within the U.S. Attorney's office whether there was any legal basis to hold Ressam. Immigration officials pointed out that there were some documentation issues that needed to be sorted out, and were able to stall a U.S. Attorney's determination that he be released. In the interim, the agents rang up the state of Washington's bomb inspection team, who arrived several hours later and confirmed that the materials in the trunk were indeed explosives.

Inspector Dean certainly deserves the nation's gratitude for having triggered the chain of events that led to Ressam's arrest. We should be thankful that the customs and immigration supervisors in Port Angeles resisted the normal bureaucratic tendency *not* to act when in doubt. But the only conclusion that people in Washington, D.C., should have drawn from narrowly dodging the bullet is that we needed to take a close look at our level of preparedness to deal with the terrorist threat to the U.S. Instead, once the millennium celebrations were over, the White House and Congress basically wiped their brows and shifted their attention to other matters.

Even after 9/11, the idea that the federal government needed to reorganize in order to address the homeland security imperative did not gather traction right away. For nine months, the Bush Administration publicly maintained that reorganization was unnecessary. But in the spring of 2002, a very small group of senior advisers began to meet in a series of closed-door White House meetings to sketch out a plan to create the Department of Homeland Security. Then, in a move that took Washington by surprise, on June 6, 2002, President Bush announced in a nationally televised speech from the Oval Office his intention to ask Congress to enact legislation that would create the new cabinet department. Congress, with some difficulty, complied. Tom Ridge, a former governor of Pennsylvania who served as the President's top advisor on homeland security, became its first secretary.

There are four compelling justifications for creating the new department. First is to bolster the operational capacity of the frontline agencies, individually and collectively, to perform their missions by correcting the kinds of problems illustrated by Ressam's arrest and my Coast Guard experiences. Second is to improve the way the federal government collects and distributes sensitive information about possible terrorist threats; the connecting-the-dots problem that marred the ability to detect and intercept the 9/11 hijackers. Third is to enhance the nation's capacity to respond to terrorist attacks. Last is to ensure better oversight by the White House and Congress of our federal security effort by putting a cabinet secretary in charge.

When the Department of Homeland Security celebrated its first anniversary in March 2004, none of these reorganization goals had been met. This is not the fault of Secretary Ridge or the other top officials at DHS. No one in the U.S. government was working harder than the team of people gathered in the cramped office space of the department's new home. But speedy results were never in the cards.

Private-sector management consultants know, from long experience in dealing with corporate mergers, that for the first eighteen months after a reorganization effort begins, costs generally go up, performance declines, and experienced people leave. The public-sector hurdles for achieving quick results are even greater. The nation's last big attempt to create and shuffle agencies under a new department offers a sobering reminder of how hard it is to merge bureaucracies and change organizational cultures. Most observers acknowledge that it took four decades after the National Security Act of 1947 established the Department of Defense to get the armed services to start working well with one another, and it is still a work in progress.

While any government reorganization effort promises to be a bumpy ride, the new Department of Homeland Security has had to face a number of start-up hurdles that go beyond run-of-the-mill bureaucratic intransigence. One handicap has been that the department needs to have lots of people with new skills. Today, immigration agents have to be cross-trained in customs laws and inspection protocols. Frontline officers must work with sophisticated technologies and be able to interact with a variety of different people, including those in the private sector. Inspectors are being sent overseas to interact with their counterparts as a part of a broad strategic effort to push America's borders out. There is the need for greater community relations to explain why new measures are being taken and to encourage the public to support those changes. These kinds of skills cannot be developed by relying on the long-standing practice of providing on-the-job training in the field. As a result, meeting the department's new mandate puts an enormous strain on rickety agency personnel systems that were built for simpler jobs in a simpler time.

Beyond the management challenges associated with identifying and meeting new personnel needs, the department has been given agencies that are reluctant to take on ambitious initiatives out of a concern that they barely have the resources to cope with the existing ones. The pace of operations associated with homeland security, particularly when it involves raised alert levels, is generating wear and tear on equipment and people. At the same time, investments to replace obsolete facilities and systems continue to plod along at a glacial pace. For instance, while the ships and aircraft of the Coast Guard are out on more frequent and longer security patrols, the service is still operating on a pre-9/11 acquisition plan that will take up to twenty-seven years to replace its already ancient fleet.

The senior management at the new department must also wrestle with the challenges associated with intruding on the turf of the other departments in the executive branch. For example, new security rules to protect our borders raise important challenges for maintaining the free movement of people and goods. But DHS is not staffed to reach out to other departments on an ongoing basis. Nor do the State, Treasury, Commerce, Transportation, and Agriculture Departments, along with the U.S. Trade Representative, have senior people assigned to focus on the diplomatic, economic, and trade dimensions of dealing with our homeland security agenda. Inevitably, clashes among competing U.S. interests that could have been anticipated and minimized by good upfront coordination turn into bureaucratic brush fires that consume the time and energy of top officials who must endeavor to extinguish them.

One particularly gray area that DHS must sort out is how to interact with the Department of Defense. The Pentagon has been keen to maintain its autonomy by assigning to itself the mission of "homeland defense," which it defines as involving terrorist attacks that emanate only from outside the United States. Relying on this definition, defense planners have essentially found a way to carve out a niche where the armed forces patrol air space and the high seas, and prepare to respond to catastrophic attacks when they happen. While there are some liaison officers assigned to one another's staffs, by and large, the Pentagon, through its new Office of Homeland Defense and Northern Command, is marshalling its considerable expertise and resources to do its own thing.

DHS must also contend with daunting problems associated with trying to gather and distribute the intelligence to support its mission. Historically, intelligence and investigative agencies are unwilling to share classified or sensitive information with bureaucracies that they consider potential competitors for their mission. Then there is the unwritten rule in the intelligence community that you have to give something to get something. The Secretary of Homeland Security has very little to give because DHS has been assigned only the job of analyzing intelligence, not collecting it. One indication of how the intelligence role at DHS is viewed within the federal government is that more than a dozen highly qualified candidates turned down the offer to be the department's intelligence chief. Since DHS exercises no control over the FBI, CIA, and the other federal intelligence agencies, the department has no leverage by which to help shape intelligence collection priorities. As a result, it must settle for what it is given, which invariably is not enough to satisfy the pleadings for better threat information by local, state, and private sector authorities.

Even within the new department there are intimidating technical and procedural barriers to sharing information. One agency is often unable to pass along data to another agency because it is saddled with technology that makes that process very cumbersome. DHS inherited a smorgasbord of old mainframe computer systems that were designed not so much for managing data as for collecting and storing it. There are also legal barriers that must be worked out to get these systems to talk to one another. Congress has always been wary of concentrating too much information at the fingertips of government bureaucrats and has stipulated in statute just what an agency may do with the data it collects.

In short, the internal and intramural struggles the Department of Homeland Security faces within the executive branch are considerable. But Congress presents yet another enormous challenge to the department's ability to focus on its mission because of unresolved issues related to overlapping jurisdictions. In the first ten months of the new department's life, its senior officials testified on 162 occasions in front of twenty-two full committees and thirty-seven subcommittees. Since each committee hearing typically requires twenty-four hours to prepare, the department spent an equivalent of 486 work days to support appearances before the House and Senate. Back at the office, DHS officials had to respond to more than 1,600 letters from members asking for additional information within the same time period. No one would expect Alan Greenspan, the chairman of the Federal Reserve, to report to dozens of committees. Nor would we expect George Tenet, the director of Central Intelligence, and his senior managers to be spending much of their working week testifying before congressional committees.

In theory, only four congressional committees are supposed to be in charge of the homeland security mission: the Senate Committee on Governmental Affairs, the House Select Committee on Homeland Security, and the Senate and House Appropriations committees. The Governmental Affairs Committee has been assigned the homeland security mandate as part of its duties in overseeing the organization of the executive branch. Because of the small size of the Senate, this appears to be working reasonably well. However, in the House of Representatives, the work of the Select Committee is complicated by the fact that it has fifty members, most of whom are powerful chairmen of standing committees. While the Select Committee has very capable leadership in its chairman, Rep. Christopher Cox, (R-CA) and its ranking member Rep. Jim Turner (D-TX), the remaining members seem more intent on ensuring that the new committee does not encroach on their turf, rather than working toward the special committee's mandate. The result is that DHS officials end up practically living on Capitol Hill, responding to legislative inquiries which obviously detracts from the time they could spend doing their primary jobs.

The bottom line is that it will be some time before we can expect Washington's new homeland security bureaucracy to be firing on all cylinders. Given the urgency of the threat, this is worrisome. However, it is not the most pressing issue that confronts us. The bigger challenge lies with identifying how to formally engage the broader civil society and the private sector, not just the federal government, in a national effort to make America a less attractive terrorist target. This is a task that requires a much different kind of institutional framework than one that involves reshuffling agencies within the executive branch. We need to mobilize the homeland by drawing for inspiration on existing entities and practices that work.

Randolph Lerner, the chairman of the bank holding company, the MBNA Corporation, has suggested that what the U.S. needs is a homeland security framework that resembles the organizational protocols and functions of the Federal Reserve System. Lerner belongs to a group of private sector leaders from major U.S. corporations who have come together to volunteer their time and expertise to support the government's efforts to combat terrorism. In the process, the group has discovered that the U.S. government has no institutional mechanism that allows it to work in real partnership with the private sector to develop contingencies in preventing and responding to acts of terrorism.

When the Federal Reserve System was created in 1913, it was organized around the notion that effective oversight of the financial sector requires drawing on the expertise of private representatives within that sector. Additionally, the Fed's charter recognized the value of taking into account the country's diversity by creating twelve regional banking districts and establishing twenty-five branches. This structure is not purely hierarchical. The regional banks are essential to the process of collecting information about the conditions at the local level, and they provide a pool of advisors to inform national policy-making functions. Importantly, the Fed also retains a degree of independence from the federal cabinet departments. While it regularly meets and supports the work of federal agencies with specific statutory responsibilities, the Fed reports directly to the Congress, and its work is audited by the General Accounting Office.

In creating a Federal Security Reserve System, we should roughly replicate the Fed's model by creating a board of governors on the national level, ten regional Federal Security Districts, and ninety-two local branches called Metropolitan

Anti-Terrorism Committees. This system would draw on a public policy rationale similar to that which inspired the Federal Reserve in the first place. The overriding interest then was to find a way to lower the risk of serious disruptions to financial markets that led to the severe economic downturns that plagued a rapidly industrializing nation. In constructing a Federal Security Reserve System, the objective would be to develop self-funding mechanisms that more fully engage a broad cross section of our society to protect our critical foundations and to limit the risk of widespread disruption arising from terrorist attacks.

The national board of governors for the Federal Security Reserve System would play an oversight role in establishing prevention and response guidelines, and monitoring compliance with security mandates within the water, food, chemical, energy, financial, information, transportation, emergency response, and public health sectors. Its responsibility would include issuing regulations with which to carry out major federal laws governing the security of these sectors. Members of the board would be available to meet with the president's homeland security advisors and to testify before Congress. The Board would submit an annual report to Congress on the state of security within each sector, along with a national vulnerability assessment. The board would also have a role in international bodies such as the International Standards Organization and the World Trade Organization in advancing universal security standards for critical networks that span international boundaries. For technical and scientific support, the board should establish a formal relationship with the congressionally chartered National Academy of Sciences.

The board should have a chairman who would be appointed by the president and confirmed by the Senate for a five-year term. Three governors should each be appointed by the president, the House, and the Senate and given ten-year terms. The Department of Homeland Security would continue to perform its functions within the executive branch just as the Treasury Department operates alongside the Fed. The board of governors would be supported by a Washington-based staff. This staff would be responsible for providing the board with independent analyses of domestic and international issues as they relate to preventing and responding to attacks on the nonmilitary networks that support our daily lives.

The staff of the board would be charged with publishing detailed information about the Federal Security Reserve System's activities, including documenting best practices and identifying clearly defined standards for emergency preparedness. These publications would be assembled with the support of regional Federal Security Districts.

Each of the ten Federal Security Districts should be given lead responsibility for a specific critical sector, based on their relative importance within their jurisdiction. For instance, the Northeast District might be assigned the lead on financial security, the Southwest district could be given the lead on energy security, and the West Coast District might have information security as its charge. The district assigned primary responsibility for a sector would chair committees made up of representatives from the remaining districts. In the case of the districts that run along the U.S.-Canada border, they should include Canadian representatives within their ranks. The infrastructure within the border region, particularly in the energy, transportation, and agricultural sectors, are so interconnected as to make the national border between the two countries functionally meaningless. In addition, all the primary districts would be responsible for hosting an international advisory committee that includes expert private and public participants from Europe, Canada, Mexico, Japan, and other allies of the United States.

The chair of each regional district would be nominated by the president and approved by the Senate. The regional board of directors would be made up of twenty-seven members appointed by the state governors within the region. Fourteen of these would be chosen from the private sector representing each of the critical infrastructure sectors with a seat reserved for a labor union representative. Five would be chosen from the public health, public safety, and nongovernmental sectors, including a designated civil liberties advocate. Another five members would come from the academic and media communities. The federal regional director from the Department of Homeland Security would serve as co–vice chair. The Department of Justice would be allowed to select one U.S. Attorney within the district to serve as the other vice chair. The Department of Defense's Northern Command, which has responsibility for protecting the U.S. territory from armed attacks, would be allowed an ex officio and nonvoting seat on the board. With the exception of the chair and vice chairs who would serve for four-year terms, all the voting members would serve for a nonrenewable six-year term.

The districts would all be assigned support staffs whose composition would be divided between full-time public-sector employees and industry experts nominated by the private sector. These industry experts would be given a two-year leave of absence by their employers to support work at the district level. As in the Federal Reserve System, these private sector appointments would be highly selective opportunities for talented mid-level executives to better understand and to help inform the policy environment that will affect their respective sectors. All of the private participants would be provided with government security clearances.

An additional responsibility of the regional chair, with input from the co–vice chairs, would be to set the terrorism alert conditions within the district, tailored to specific sectors and locations whenever possible. This would help to temper the issue of vagueness, high-cost, and risk of growing public complacency that has marred the operation of the national five-color terrorism index. The regional district would also be responsible for conducting regional response drills and documenting gaps in capabilities and lessons learned.

At the metropolitan level, the Federal Security Reserve System could be tied into an expanded version of the post–9/11 Anti-Terrorism Advisory Committees and Joint Terrorism Task Forces that are run by the Justice Department and the FBI. Currently these groups are made up of representatives from federal, state, and local law enforcement agencies. They are essentially forums where cops can talk with other cops. Under the Federal Security Reserve System, these organizations would be set up in all states and within the major metropolitan areas and should include private sector representatives. They should be co-chaired by the local U.S. Attorney and a person with private sector background who is appointed by the regional district.

The Metropolitan Anti-Terrorism Committees should serve as a forum for sharing the latest threat assessment information with local and private entities that have direct operational oversight over critical sectors. This would help to address one of the major sources of frustration among local authorities and private sector leaders, that the federal government does not share its intelligence with them. It would also help the federal government by tapping the insights and expertise of industry experts and local police, which can be very useful in making sense of the intelligence "chatter" that federal agencies receive each day. To satisfy security concerns over control of sensitive information, the integrity of the private sector participants would be vouched for by the companies they come from and they could be subjected to the same background checks used for federal employees.

Finally, the metropolitan committees would be charged with reviewing vulnerability assessments and security plans, including conducting red-team exercises to evaluate the on-the-ground reality of prevention and protection efforts. The teams would also be given responsibility from time to time to inspect compliance with security regulations in much the same way that federal bank examiners are sent out from time to time assess the operations of their member banks. When serious discrepancies are found, the district board could impose sanctions.

The overall thrust of this proposed Federal Security Reserve System is to create a participatory system that does not place unrealistic reliance on the activities of federal agencies. By using the Federal Reserve System as a template for enlisting expertise beyond Washington, we can achieve a middle ground between placing the fate of critical networks entirely in the hands of overworked federal authorities, and relying on an unworkable laissez-faire approach that provides no protection.

The Federal Reserve describes itself as a system that is "independent within the government." This means it must work within the overall objectives established in law by Congress, but its decisions do not have to be ratified by the president or anyone else in the executive branch. This level of independence is justified as being both a check on executive power and as a way to manage the risk that decisions directly affecting the operation of the marketplace might become dangerously politicized. The case for a similar approach to homeland security is a compelling one. In both instances, the goal is to better align commercial interests with public interests.

The downside of an independent agency is that it can become co-opted by the very private sector elements that it is supposed to regulate. That risk tends to rise when the agency operates largely out of the public eye, or when the public interests it is responsible for safeguarding are not deeply embraced. But the proposed Federal Security Reserve System would have a high profile and its mandate would be to protect a cherished public good. The diversity of its membership also promises to reduce the risk that advocates for short-term market benefits would be allowed to carry the day.

Another way to help reduce the risk that the homeland security agenda could fall victim to partisan politics is to provide the Federal Security Reserve System with an independent source of funding. The way to accomplish this is to create a catastrophic terrorism trust fund that is bankrolled by fees levied on critical infrastructure activities. For example, fees could be assessed on the movement of containers, water and electricity usage, refinery production, the movement of natural gas or oil through a pipeline, and at the gas pump to pay for bridge and tunnel protection.

The fund would have two purposes. One would be to provide a day-after fund to make resources immediately available in response to an act of terror and to repair the damage to the systems that were targeted. The other is to use the interest earned on the fund to pay for the overhead costs associated with running the Federal Security Reserve System.

This trust fund approach would not be new. After the 1989 grounding of the supertanker *Exxon Valdez* in Valdez, Alaska, an Oil Spill Liability Trust Fund was set up by Congress the following year. This one-billion-dollar fund was financed by imposing a fee of five cents per barrel on domestically produced and imported oil. Once the fund was financed to its authorized level, the fee was suspended. If the fund is drawn upon in the aftermath of a major oil spill, the fee could automatically be reactivated to replenish it. The fund, which grew with accumulated interest, can also be used by Congress to pay for various oil-safety and pollution-prevention initiatives.

There is another precedent that could be borrowed from the oil pollution model. It is to mandate that owners and operators of critical infrastructure carry adequate levels of insurance. This not only reduces the call on public resources when acts of terror occur. Importantly, it also creates incentives for the insurance industry to become a partner in ensuring that the owner and operators of essential systems do not neglect their security responsibilities.

In the case of the Oil Pollution Act of 1990, oil tankers are required to provide evidence that they carry a minimal level of insurance, based on their size, to cover the costs associated with an oil spill, should they have an accident. This evidence in the form of a certificate of financial responsibility (COFR) must be on file with the U.S. Coast Guard as a condition of gaining entry into U.S. waters. Beyond this minimum level, the Oil Liability Trust Fund is available to pay victims who might otherwise not receive compensation. To reduce the risk of complacency, the Oil

Pollution Act includes a clause that lifts the liability ceiling if there is evidence of gross negligence, willful misconduct, or violations of the federal regulations. Since insurers are always interested in reducing their exposure to large claims, many carefully inspect vessels to ensure shipowners are abiding by the prescribed standards and deny insurance to those they find are not up to snuff.

The Federal Security Reserve System board could address the liability issues in its sphere by establishing minimum levels of insurance for each critical infrastructure sector and require as proof of insurance a certificate of security responsibility. The catastrophic terrorism trust fund would be used when the costs associated with a terrorist attack exceed that liability level. As with the oil tanker industry, the liability level could be removed if there is evidence that negligence contributed to the losses associated with the attack. To receive insurance, prevention and response plans would be reviewed by the insurer. The metropolitan counterterrorism teams would periodically examine and test these plans to create an additional incentive for compliance. If discrepancies are discovered, the rating of the insurers could be lowered. Also, since the U.S. government is acting as an insurer of last resort, it should assess a fee on all insurance policies, which can be used to help defray the overhead costs associated with operating the metropolitan counterterrorism teams.

The Oil Pollution Act also provides a Federal Reserve-like organizational structure that demonstrates that a collaborative private-public model can work for more than just finance. The act authorizes the creation of area committees at the regional level to bring industry and public sector representatives together to develop contingency plans for managing major spills. This planning process compels everyone involved in the aftermath of a pollution incident to talk with one another in advance. It also helps to ensure that there will be an organized process for managing oil spills, thereby resolving the ad hoc, "come as you are" approach that had marred the response to earlier oil pollution incidents.

In the area of oil pollution, this framework has a proven record of success that should evoke enthusiasm for applying it to homeland security. It has worked as a deterrent for those ships that have historically operated on the margins of safety that now rarely attempt to enter U.S. ports. Today, oil tankers that do enter U.S. waters are now operating more safely as evidenced by a 95-percent drop in the volume of oil spilled from pre-*Exxon Valdez* levels. Prior to 1990, 70,000 barrels per year ended up polluting American waters. Now the annual average is four thousand barrels.

Of course, a Federal Homeland Security System would never be entirely removed from Washington's worst political impulses. But, like the Fed, it would be better than the alternatives. Built around regional districts and tapping expertise and relationships that the executive branch cannot replicate, it draws upon America's federalist character and the strength of our civil society. This system can serve another salutary purpose by forcing Congress to play a more disciplined and active role in supporting the security of the nation. Drawing on input from the ten regions, the board of governors of Federal Security Reserve System would apprise Congress each year of the needs, resources, and requirements to make the American homeland secure. It would then fall to the legislative branch to act on that information.

In reorganizing our security apparatus in order to confront the dangers of terror and asymmetric war, we should draw inspiration from the example of our Founding Fathers. The U.S. Constitution was crafted not in the immediate aftermath of the War of Independence, but years later, when it became apparent that our first attempt to organize the government under the Articles of Confederation had produced a not-so-perfect union. The men who gathered in Philadelphia in 1787 had the courage to formulate and advocate bold changes to ensure the future vitality of the experiment in democracy that they had launched with the Declaration of Independence.

We should also be mindful of the words of Abraham Lincoln, included in a message to Congress in December 1862:

> The dogmas of the quiet past are inadequate for the stormy present. The occasion is piled high with uncertainty, and we must rise to the occasion. As our case is new, so we must think anew, and act anew. We must disenthrall ourselves, and then we shall save our country.

This is not a time for timidity, nor is it a time for persisting with an outmoded national security framework designed for a different enemy in a century that has now gone by. We must create the kind of structure that widens the breadth and quality of civic participation in making America safe. The nation, not just the federal government, must be organized for the long, deadly struggle against terrorism.

8

Fear, Paranoia, and the Gathering Storm

Over the past few years, I have traveled across America speaking to audiences candidly about our national state of insecurity. Initially, I approached the enterprise with some trepidation, given that my message was unsettling. But over time it became clear that people want to hear the facts, however alarming. Even more encouraging has been the extent to which everyday people have expressed their sincere desire to be helpful in some way. Once they learn of the dangers we face, most of the Americans I meet want to contribute more to our national well-being than simply to continue shopping and traveling. They report being frustrated that the U.S. government is not tapping into their sense of civic duty.

This desire to help stands in marked contrast to the sense of skepticism, even resignation, with which the general public has greeted the periodic raising and lowering of the terror-alert levels.

These somber warnings of impending danger stand in stark relief against the backdrop of perky reassurances that vague security measures have everything under control. Especially worrisome is the extent to which the public appears willing to second-guess official directives. According to a Columbia University study released in August 2003, 90 percent of Americans polled said they would not evacuate their homes in time of a crisis based only on a government order to do so. Instead, they would seek out additional information before deciding what to do. Public health officials are increasingly worried that weak public confidence in official guidance will undermine their efforts to carry out mass vaccinations and impose quarantines in the face not only of bioterrorist attacks but also of outbreaks of SARS and other naturally occurring diseases.

Washington should be working overtime to stem this erosion in public trust. But the federal government's tendency to deal behind closed doors with all security matters is only fueling the problem. As Senator Warren Rudman has warned, "There will be hell to pay" when the quiet decisions to postpone the response to known vulnerabilities become public knowledge in the aftermath of the next attack. Americans will be rightfully enraged to learn that senior government officials were aware of the threat but had concluded that putting adequate safeguards in place to protect their citizens was too difficult or too expensive, and then hid from the electorate both the reality of the danger and the decision not to do much about it. The likely justification for keeping us in the dark will be that any public disclosure of the decision to live with our vulnerabilities would have been imprudent because it would advertise our weakness to our enemies. Such a defense should be dismissed as specious.

"We the people" are the first three words of the preamble of the U.S. Constitution. "To provide for the common defense" is then listed among the reasons for forming a more perfect union. If our senior government officials believe that they cannot credibly provide for that defense, they have a constitutional obligation to level with us. It is then up to us to decide if we are prepared to draw on the resources of our civil society to mobilize a response.

Citizenship carries responsibilities, which include contributing more than just tax dollars to the cause of protecting our way of life. In times of national emergency we have an obligation to play a more active role. This principle is well grounded in our history, dating back to the Minutemen, who traded in plowshares for muskets so they could fight in the American Revolution. It is what helped transform America from a second-rate military power into a victorious force during World War II. At the start of that war, Japan and Germany saw us as a complacent society, so consumed by a self-centered pursuit of our material well-being that we would be unwilling and unable to stand up to their aggression. They badly miscalculated.

However, the practice of incorporating civil society to provide for the common defense waned during the Cold War era. With the Soviet Union, we faced an adversary who was prepared to launch thousands of nuclear warheads on U.S. cities within a matter of minutes. Given such a clear and present threat, there was little choice but to place our defense in the hands of a massive national security establishment that maintained around-the-clock vigilance. With the risk of nuclear Armageddon being managed in the background, we got into the habit of expecting Washington to protect us from any and all threats posed by our enemies.

The federal government, in turn, grew increasingly accustomed to playing a paternalistic role when it came to our security. Perversely, in many quarters ordinary citizens were viewed as part of the problem rather than the solution. Security officials often act as though members of the American public are either potential recruits for an easily panicked mob or a passive part of a haystack that must constantly be sifted to find terrorist needles. This perspective is reinforced by a government apparatus that operates on the basis that national security information should only be shared with those who have a "need to know." Even within the government, the only people who are allowed access to details of national security are those with the appropriate clearances and specific duties that require it to do their jobs.

The requirement of "need to know" is appropriate for issues surrounding nuclear weapons and the technical details associated with intelligence collection, communications systems, and military hardware used in waging conventional war. But the new threat environment requires that our government refrain from keeping its security cards close to its chest. Rather than working assiduously to keep the details of terrorism and our vulnerabilities out of the public domain, the federal government should adopt a new imperative that recognizes that Americans have to be far better informed about the dangers that they face.

In short, the U.S. government needs to stop treating the American people as though they are children. The contention that openly talking about our myriad vulnerabilities will only give our enemies ideas is overdrawn. For one thing, it grossly underestimates the determination and capability of our adversaries. It also underrates what can be quickly accomplished by harnessing the ingenuity and civic-mindedness of the American people.

One has only to look to New York City on the day of September 11, 2001, to make the case that everyday citizens are up to the task of playing an active role in managing the threats that now confront our nation. In those first critical hours after the World Trade Center towers came down, the federal government and its security machinery was nowhere in sight. Local officials and the people of New York responded heroically, in striking contrast to the heinous acts that had been visited upon them.

Those people living and working in America's biggest city did not turn into a mindless mob in the face of a calamitous disaster. Firefighters and police officers sacrificed their lives to try to save perfect strangers, and people committed countless acts of kindness and bravery. Everyone pulled together to evacuate Lower Manhattan, whether as a part of the spontaneous armada of waterfront vessels that showed up to ferry people away by sea or by storeowners handing out new shoes, clothes, and free drinks to the people who hiked north to take trains out of the city. And then, without anyone asking, construction workers started showing up with their heavy equipment, or just their bare hands, volunteering to join the emergency responders in the search for life amid the devastation of Ground Zero.

When the mass electrical outage turned off the lights in New York City on August 14, 2003, we once again witnessed that the public did not panic. Nor were there mass lootings on the street. Instead, neighbors helped neighbors who were trapped in apartment elevators, assisted one another in climbing out of pitch-black subway tunnels, and helped direct traffic on the streets of Manhattan so that emergency vehicles could get through crowded intersections. During the outage, the overall city crime index was lower than it was for the same period in the previous year.

Michael Bloomberg, the mayor of New York, received high public approval ratings for his leadership during the outage. But two months later, Bloomberg did something very un-Washington-like. He publicly released a lessons-learned report that exposed shortcomings in the emergency response to the crisis. Bloomberg recognized that things could have turned out much worse if the outage had coincided with a terrorist attack that involved a large number of casualties. He also understood that resolving shortcomings would require a supportive public. While releasing the report might cause commentators to reconsider the praise they had showered on the mayor, Bloomberg decided that the real risk lay in not acknowledging what needed to be fixed before the city faced the next major public safety crisis. He accepted the fact that there almost certainly will be another incident, probably sooner than later.

In short, everyday Americans are likely to be more resilient in times of crisis than our elected leaders assume. Still, Americans should be more wary about our political system's capacity to behave reasonably in the aftermath of these crises. A major terrorist attack brings with it enormous pressure for elected leaders to embrace drastic measures to confront the revealed danger. This dynamic has, as its driving force, a desire to be seen as providing decisive leadership in times of crisis. By seizing upon radical, even draconian, new security initiatives, elected officials convey the image of being in charge during time of national peril.

Politicians in the minority party rarely rally against this post-crisis tendency to allow the security pendulum to swing too far. Indeed, their instincts are often to start a bidding war in an effort to convey that the public would have been better protected had they been in the majority. The result is that the voices that might call for debate are largely mute, and the mad dash to create new security laws becomes a bipartisan enterprise. "Act now" becomes the mantra, rather than "act effectively."

In short, the checks and balances that were so skillfully incorporated into our constitutional framework are weakest in times of crisis. There is no guarantee that Congress and the White House will do the right thing after the next attack. Indeed, historically, our liberties have been most vulnerable during and immediately following a national trauma.

Americans would do well to reflect on how a lack of preparedness can backfire on our most cherished values by recalling the World War II decision to place 120,000 citizens and resident aliens of Japanese descent into detention camps in the Rocky Mountain region. After the attacks on Pearl Harbor, evacuees were forced to leave their homes and most of their belongings, and they remained in these remote camps until January 1945. Italian-Americans were also detained. Former CIA director James Woolsey has argued that what makes this story a particularly poignant cautionary tale is not that the order to intern Japanese-Americans was made by an overzealous and racist U.S. Army commander charged with defending the West Coast from a Japanese attack. Rather this decision was supported by President Franklin D. Roosevelt; by Earl Warren, then governor of California and later the chief justice of the Supreme Court; and by Hugo Black, the Supreme Court justice who delivered the highest court's opinion upholding the internment in *Toyosaburo Korematsu* v. *United States*. In other words, this gross injustice was accepted as a necessary exigency of war by three of the liberal titans of the twentieth century.

Six decades later, there can be no guarantee that mobilizing the nation to prepare for the gathering storm of the new shadow warfare will eliminate the risk of a wavering commitment to our democratic values. Depending on how successful our enemies are in carrying out the next attack on U.S. soil, the political pressure to embrace security-at-any-cost measures rather than for effective risk management measures may prove overwhelming. What is certain is that the risk of overreach rises if government officials are charged with having done too little, too late. This means that "we the people" should consider an early and substantial investment in prevention, protection, and response measures as a form of insurance against self-inflicted harm to our liberties and way of life.

How much security is enough? We have done enough when the American people can conclude that a future attack on U.S. soil will be an exceptional event that does not require wholesale changes to how we go about our lives. This means they should be confident that the measures in place are sufficient to confront the danger.

Homeland security should strive to achieve what the aviation industry has when it comes to safety. What sustains air travel despite the periodic horror of airplanes falling out of the sky is the extent to which long-standing and ongoing investments have convinced the public that it is safe to fly. Public confidence never can be taken for granted after a major jetliner crash, but private and public aviation officials start from a credible foundation built upon a cooperative effort to incorporate safety into every part of the industry. Every time we board a plane, we receive instructions from the flight attendants on how to fasten our seatbelts and don oxygen masks as a gentle reminder of the paramount importance that the industry assigns to safety. In the immediate aftermath of airline disasters, the public is reassured by the fact that the lessons learned are quickly compiled and released and that their government and the industry seem willing to take whatever corrective actions are required.

The aviation-safety example highlights the extent to which ongoing and credible efforts to confront risk are essential to the viability of any complex modern enterprise. Similarly, in the face of post–9/11 threats, the viability of our cherished way of life is dependent on quickly developing and maintaining a credible means of securing our homeland. In undertaking this endeavor, we should be guided by seven principles:

There is no such thing as fail-safe security, and any attempt to achieve it will be counterproductive. Americans were not perfectly secure before 9/11, and they will never be perfectly secure. We can no more hope to defeat terrorism completely than we can expect to eliminate crime or the influenza virus. Ultimately, our goal must be to manage the risk of acts of terrorism on U.S. soil. While we should aspire to deter or foil enemy designs to destroy American lives and property, we don't want security to become such an overarching preoccupation that we sacrifice our democratic ideals. In the end, the cost of security protocols cannot be so high as to bankrupt the very things we are trying to secure.

Security must always be a work in progress. If our security practices become ritualized, they become fixed targets that our enemies can compromise or evade. Groups like al Qaeda have demonstrated that they can be perversely creative, patient, and tenacious. They invest in staking out targets and probing for weaknesses. Thus, security protocols should constantly be adjusted to keep the bad guys guessing and to keep pace with any changes to the system the protocols have been crafted to protect. We must make a concerted effort to anticipate the most likely future-attack scenarios and create protective measures in advance. These measures must form a layered and dynamic system of defenses that incorporate the contribution of human intuition and judgment.

Homeland security requires forging and sustaining new partnerships at home and abroad. Reducing the vulnerabilities that terrorists are most likely to exploit or target involves harnessing the expertise that lies mostly outside the federal bureaucracy. Most of the critical infrastructures that terrorists might want to destroy or disrupt are linked to global networks, so this means that including citizens, companies, and countries in any new security regime is vital. The secretive, top-down, us-versus-them culture that is pervasive in government security circles must give way to more inclusive processes. It also means that new federal security measures cannot be developed in a vacuum. At a minimum, they must be evaluated against the risk of becoming a source of angst and alienation within ethnic and business communities, among our trade partners, and in the public at large. For instance, if strong-armed enforcement tactics are perceived as unfair, then the general public will be less willing to report suspicious activities and our allies will be reluctant to share information or to cooperate in counterterrorism operations.

Our federalist system of government is a major asset. Given the size and complexity of American society, there are no one-size-fits-all approaches to addressing the nation's most serious homeland vulnerabilities. Private-sector leaders and local officials who are most familiar with those vulnerabilities will generally have the best insights on effective solutions. National coordination, resource support, and leadership by the federal government are essential. Encouraging regions, states, localities, and the private sector to experiment and to adapt to local and regional circumstances will ensure that our nation's approach to homeland security will be as dynamic as the threat that confronts us.

Emergency preparedness can save lives and significantly reduce the economic consequences of terrorist attacks. During the Cold War, most Americans believed that civil defense measures were futile. Today's security environment mandates that we put this sense of resignation behind us. An attack with a nuclear, chemical, or biological weapon will affect not only those who are immediately exposed but could also wipe out the emergency response and medical care system in the surrounding areas. Heavy losses of seasoned firefighters, emergency technicians, police, and medical personnel can easily compromise a community's long-term capacity to provide for public health and safety. If the effects of the attacks are not quickly contained, the cleanup costs, particularly in a busy urban area, are likely to be prohibitively high.

Homeland security measures have deterrence value. While terrorist acts will remain a fact of modern life, we should not allow them to be the military tactic of choice for confronting U.S. power. The attraction of targeting our civil society and economic assets is its potential for compelling the U.S. to reconsider the costs of policies or actions that our enemies find objectionable. If these attacks are likely to be detected, intercepted, contained, and managed without any serious fallout, then their military usefulness will have to be reevaluated. Since these acts violate widely accepted norms, they will almost certainly invite not just American, but international retribution. Most adversaries would probably judge this too high a price to pay if the acts do not produce the desired mass disruption.

Homeland security measures will have derivative benefits for other public and private goods. Most of the measures enacted to prevent and mitigate the consequences of a terrorist attack will offer additional benefits. Bolstering the tools used to detect and intercept terrorists will enhance the means that authorities have to combat criminal acts such as narcotics and migrant smuggling, cargo theft, and violations of export controls. Diseases such as SARS, AIDS, West Nile, foot-and-mouth, and mad cow have highlighted the challenges of managing deadly pathogens in a shrinking world. Public health investments to deal with biological agents or attacks on our food and water supplies will provide U.S. authorities with more effective tools to manage these global diseases and pandemics. The measures we adopt to protect our infrastructure make it more resilient not only to terrorist attacks but to acts of God or human and mechanical error.

Ideally, this last principle should help us overcome some of the widespread ambivalence about tackling the homeland security agenda. Managing the risk of terror need not involve costs with no associated benefits. Often the opposite is true. This means that sustaining public support for confronting the new security environment should be politically feasible, even in the absence of another 9/11-style attack.

* * *

Many times in recent years, particularly since 9/11, I have had to fight back a sense of despair. For most of my adult life I have been acquiring an intimate understanding of America's vulnerabilities; I have been schooled in the motives and capacities of our enemies; I have lobbied for change in government and industry to better manage these risks; and I have borne direct witness to the painfully slow, and at times ill-conceived, ways that Washington has been approaching our homeland security challenge. These experiences have conspired to fill me with a sense of dread that we continue to live on borrowed time, where the likelihood of another deadly attack on U.S. soil rests more with the designs of our enemies than with the means that our government has been cobbling together to protect the U.S. I have been frustrated by the sense of denial that pervades so many corners of the federal government. There is a dangerous proclivity among U.S. officials to believe their own rhetoric about how much progress they have been making on issues that require reversing decades of neglect.

Yet I remain optimistic about the future. Leadership begins with acquiring an unvarnished view of the realities that define current circumstances. Then it requires acting on the conviction that those realities can be transformed. America's security and prosperity have never rested so much on geography as they have on our historic willingness to embrace and make sacrifices for our ideals. For many Americans, those ideals became more precious after the horror of September 11. We now need to reclaim them, not just for ourselves but as the ultimate weapon in fighting the global war on terrorism. Like those who led America in the aftermath of World War II, we must commit ourselves to engaging the world around us, bolstering the institutional capacity to manage our common problems, and demonstrating the generosity of the American spirit. We must continue to remind the world that it is not military might that is the source of our strength but our belief that mankind can govern itself in such a way as to secure the blessings of liberty.

Notes

1 Living on Borrowed Time

Page

2 *While we receive a steady diet:* "The Homeland Security bill I will sign today commits $31 billion to securing our nation, over $14 billion more than pre-September 11th levels." President George W. Bush. Remarks made by the President on signing the Homeland Security Appropriations Act, Department of Homeland Security, Washington, D.C., October 1, 2003. Total Pentagon spending as estimated by the Center for Defense Information in 2004 is $476 billion dollars.

4 *At one end of the spectrum:* "In Mozambique and Angola, an AK-47 rifle, complete with two clips of ammunition, can be purchased for less than $15 or a bag of corn." United Nations Regional Center for Peace and Disarmament in Africa, "Small Arms Transparency and Control Regime (SATCRA) Program," www.unrec.org/eng/SAT-

CRA.htm. "In Uganda . . . an AK-47 automatic rifle, of which there are around 55 million in circulation, and which can be easily carried, maintained and assembled by a ten-year-old child, can be bought for the price of a chicken and in Northern Kenya for the price of a goat." The First Prepcom for the 2001 UN Conference on Small Arms Trafficking in All Its Aspects. "The Humanitarian Challenge of Small Arms Proliferation: The argument for a universal, comprehensive and non-discriminatory small arms control regime," white paper prepared by His Excellency Mr. Alpha Oumar Konare, President of the Republic of Mali and His Excellency Mr. Michel Rocard, former Prime Minister of France, co-chairs. The Eminent Persons Group, United Nations Headquarters, New York, March 3, 2000; Warren Richey, "In gun seizures, some surprise finds." *Christian Science Monitor,* May 15, 2003. In postwar Iraq, the demand for weapons, combined with U.S. efforts to reduce the supply, have sent the price of an AK-47 from $8 to $80 since VE day.

4 *Weapons-usable nuclear materials exist:* Matthew Bunn, Anthony Wier, and John P. Holdren, *Controlling Nuclear Warheads and Materials: A Report Card and Action Plan*, Project on Managing the Atom, Commissioned by the Nuclear Threat Initiative. (Cambridge, MA: Harvard University 2003): xii, www.nti.org/e_research/cnwm/cnwm.pdf.

6 *Fewer than one hundred Romans:* Peter S. Wells, "*The Battle that Stopped Rome: Emperor Augustus, Arminius, and the Slaughter of the Legions in the Teutoburg Forest* (New York: W.W. Norton and Company, October 2003).

6 *"First, a symmetrical and overwhelming approach"*: General Charles C. Krulak, Commandant, U.S. Marine Corps. Remarks at the Council on Foreign Relations, November 17, 1997.

7 *In that manifesto, bin Laden:* The full text of the manifesto, in the original Arabic and in translation, is widely available online, including at www.islamic-news.co.uk/declaration.htm.

7 *Two years later in a joint statement:* World Islamic Front Statement, "Jihad Against Jews and Crusaders," February, 1998, www.fas.org/irp/world/para/docs/980223-fatwa.htm.

7 *They have lost this:* Rohan Gunaratna, *Inside Al Qaeda, Global Network of Terror* (New York: Columbia University Press, 2002): 226; Tayseer Allouni [the Kabul correspondent of *Al-Jazeera*], transcript of an interview with Osama bin Laden, October 21, 2001, translated from Arabic by the Institute for Islamic Studies and Research, www.alneda.com.

8 *They argue that the new principles:* Qiao Liang et al., *Unrestricted Warfare: China's Master Plan to Destroy America* (Panama City, Panama: Newsmax.com, 2002). The book was originally translated into English by the CIA before it was published commercially.

12 *In 2002, nearly 400 million people:* Passenger data: Performance and Annual Report, Fiscal Year 2002, U.S. Customs Service, p. 15; car, truck, and rail freight data,

National Transportation Statistics 2003, Bureau of Transportation Statistics; Maritime container data based on, "Top 30 U.S. Ports Calendar Year 2002," U.S. Waterborne Foreign Trade, Containerized Cargo, U.S. Maritime Administration, February 26, 2003; Vessel data, "Vessel Calls at U.S. Ports 2000," Maritime Administration, Office of Statistical and Economic Analysis, March 2004.

12 *U.S. customs and border protection inspectors:* U.S. Customs Service, Performance and Annual Report: Fiscal Year 2002: 9.

12 *And then there are the 7,000 miles:* The length of the U.S.-Canadian border excluding Alaska, is 3,987 miles. The length of the Alaska-Canadian border is 1,538 miles. The length of the U.S.-Mexican border is 1,933 miles. Based on figures provided by the U.S. Geological Survey, Department of the Interior to the *World Almanac and Book of Facts, 2001*: 619; shoreline length based on figures from NOAA Coastal Services Center, www.csc.noaa.gov/.

12 *Official estimates place the number:* The former INS puts the unauthorized immigrant population residing in the United States at 7.0 million as of January 2000. Office of Policy and Planning, U.S. Immigration and Nationalization Service, *Estimates of the Unauthorized Immigrant Population Residing in the United States: 1990 to 2000* (Washington, D.C.: January 31, 2003), uscis.gov/graphics/shared/aboutus/statistics/2000ExecSumm.pdf.

2 The Next Attack

Page

18 *Let's say his name is Omar:* Details of the Dar al `Ulum Haqqaniyah madrassas can be found in Peter Bergen, *Holy War Inc.* (New York: The Free Press, 2001): 149. The madrassa is headed by Maulana ul-Haq. (Maulana is a senior religious title) who presides over 2,800 students, mostly from Pakistan, but also including hundreds from Afghanistan and dozens from the Central Asian republics of the former Soviet Union. According to Bergen, this school educated as many as eight cabinet-level members of the Taliban.

18 *There he committed the Quran:* For a broader discussion of radical Islam see Jessica Stern, *Terror in the Name of God* (New York: Ecco, 2003); Bernard Lewis, *The Crisis of Islam* (New York: Random House, 2002); Stephen Schwartz, *The Two Faces of Islam* (New York: Anchor, 2003); and Hamid Algar, *Wahhabism: A Critical Essay*, (Islamic Publications International, March 2002).

18 *He finds work as a ship breaker:* Paul Haven, "Ship 'breaking' resurfaces in Pakistan: Dismantling has its environmental and health hazards," *Associated Press*, December 22, 2003.

19 *There he uses an al Qaeda-provided:* Jeff Goodell, "How to Fake a Passport," *New York Times Sunday Magazine*, February 10, 2003.

19 *The material is americium-241:* My decision here to use

the threat of dirty bomb attacks by my hypothetical al Qaeda cells is drawn from recent discoveries that indicate al Qaeda and other terrorist organizations have been aggressively pursuing the means for such attacks. When coalition forces invaded Afghanistan in the fall of 2001, evidence began to emerge that al Qaeda leaders had advanced plans to deploy a radioactive (dirty) bomb against U.S. or other targets. In early 2002, the *Washington Post* reported that it was the consensus view in the White House and the intelligence community that al Qaeda may be in possession of the necessary material, such as strontium 90 or cesium, to deploy a bomb. To date, however, it is Chechen rebels, not al Qaeda, who have pursued dirty bombs most rigorously and with a frightening degree of success. In 1995, Chechen rebels informed a Russian news channel that they had burried a container of cesium in Moscow's Ismailovsky Park. The container was found, but the source of the material was never discovered, nor were the perpetrators ever identified. In December 1998, Chechen security forces, aligned with the Russian government, discovered and defused a radioactive bomb in a suburban area ten miles outside the Chechen capital of Grozny. See "Dirty Bomb," NOVA, Original PBS Broadcast Date, February 23, 2003 (transcript and other information available online at www.pbs.org/wgbh/nova/dirtybomb); Barton Gellman, "Fears Prompt U.S. to Beef up Nuclear Terror Detection," *Washington Post*, March 3, 2002; and John Mintz and Susan Schmidt, "'Dirty Bomb' Was Major New Year's Worry," *Washington Post*, January 7, 2004. In this scenario, the americium would likely come from oil-well surveying equipment where it is used because it emits radioactive alpha particles as the device is lowered deep into the well to help establish whether or not there are fossil fuels.

19 *In 2002, Ukrainian law enforcement:* U.S. Department of State, *International Narcotics Control Strategy Report* (U.S. GPO, Washington, DC: 2002): IX–130.

20 *The accomplice's specialty:* "The Terrorist Within: Chapter 8, Going to Camp," *Seattle Times*, June 23 to July 7, 2002.

21 *When the bomb is detonated:* Federation of American Scientists Web site, "Testimony of Dr. Henry Kelly, President of the Federation of American Scientists before the Senate Committee on Foreign Relations, March 6, 2002," www.fas.org/ssp/docs/030602-kellytestimony.htm. Dr. Kelly's testimony is based on an analysis prepared by Michael Levi, Director of the Strategic Security Program at the Federation of American Scientists (FAS), and by Dr. Robert Nelson of Princeton University and FAS.

21 *Mannheim and neighboring Ludwigshafen:* See City of Mannheim, Germany Web site: www.tourist-mannheim.de/tourist-center/stadtrundfahrten/c_stashe.htm

21 *One of their cyber-literate:* An interview with Michael Surridge, Commissioner of Revenue Protection, Government of Jamaica, in Kingston, Jamaica, on August 16, 2000, revealed that criminal gangs have undertaken similar plots involving hacking into a shipper database. For example, this technique was used by a luxury-car theft

ring operating out of southern Florida. They hacked into a Canadian automobile exporter's database, downloaded the vehicle identification numbers (VIN) and vehicle descriptions, and then went out and stole luxury cars of the same make and color. They then fabricated counterfeit VIN plates with the stolen numbers and swapped them with the originals and shipped the vehicles to Kingston, Jamaica, pretending to be the Canadian exporter. In doing so, their plan was to sell the cars on the local market.

22 *It's almost certainly overkill:* As an indication of the small odds that a container from a trusted source will receive even a cursory physical examination, there is the example of a cargo container from a plant owned by Osram—Sylvania that was shipped to Hillsboro, New Hampshire, in a U.S. government-sponsored test of container security in May 2002. The tested container, which contained light bulbs, was outfitted with a global positioning system (GPS) device and a series of light and magnetic sensors that could detect an intrusion into the container. The container door was adorned with a six-inch GPS antenna temporarily held in position with magnets, from which ran several foot-long black wires through the door gasket to an automotive battery and sensor equipment bolted to the interior side of the container wall. Despite the fact that this equipment should have struck inspectors as unusual and potentially frightening, the container passed through five international jurisdictions and in and out of two seaports without any official ever stopping it. For more information see Stephen Flynn and Peter Hall,

"Bolstering Cargo Container Security: How U.S. Attorneys' Offices Can Make a Difference," *United States Attorneys' Bulletin* 52:1, January 2004: 35–42.

22 *Rotterdam is Europe's biggest:* Port of Rotterdam, "Port Statistics 2002," www.portofrotterdam.com/Images/16_52849.pdf.

23 *Working with their Dutch counterparts*: Interviews with Dutch customs inspectors during a field visit to the Port of Rotterdam on February 6, 2001, and interviews with U.S. customs inspectors involved in the Container Security Initiative program.

23 *The vessel is manned:* In an effort to boost profits in the face of fierce competition, ship owners are operating larger and larger vessels with smaller and smaller crews. In 1990, *New Yorker* essayist John McPhee wrote about the decline of the U.S. merchant marine and the push to trim crew sizes and build larger vessels in *Looking for a Ship* (New York: Farrar Straus & Giroux, 1990).

24 *Hassan left Algeria in 1993:* The character of Hassan is loosely based on the millennium bomber, Ahmed Ressam. For details of Ressam's life and terrorist activity, see "The Terrorist Within," *Seattle Times*, June 23–July 7, 2002.

24 *The fact that organized crime:* Interviews with Royal Canadian Mounted Police and Revenue Canada officials in Montreal on August 14, 2001.

24 *The radioactivity of cesium-137:* In February of 2002, a lost medical gauge containing cesium was found in

North Carolina. Other such devices are widely used in the medical field and in industry. See the Federation of American Scientists Web site, "Testimony of Dr. Henry Kelly, President of the Federation of American Scientists before the Senate Committee on Foreign Relations, March 6, 2002," www.fas.org/ssp/docs/030602-kellytestimony.htm.

25 *By the end of 2001, there were five thousand:* Charles D. Ferguson et al., *Commercial Radioactive Sources: Surveying the Security Risks*, The Center for Non-Proliferation Studies, Occasional Paper No. 11 (Monterey, CA; Monterey Institute of International Studies: January 2003).

25 *The tribe owns over 30,000 acres:* A phone conversation with Jennifer Jock, Public Information Director for the Saint Regis Mohawk Tribe, revealed that the U.S. portion of Akwesasne consists of 13,422 acres that is federally recognized and 2,178 acres which include two parcels of disputed property, totaling 15,600 acres of land. The Mohawk Council of Akwesasne on the Canadian side has approximately 22,000 acres.

25 *Back in the days of Prohibition:* Use of the reservation as a smuggling route has been well-documented. For example, see Paul Zieblauer, "26 Are Arrested in Drug Smuggling Network Using Mohawk Reservation," *New York Times*, June 9, 1999.

26 *Eight thousand pounds of ammonium nitrate:* The bomb that Timothy McVeigh built in the back of a rented Ryder truck is believed to have been made of between 4,500 and

5,000 pounds of fertilizer with a TNT booster. See James Kitfield, "Devastation in a Truck," *The National Journal* (August 15, 1998).

26 *Every day, these rigs:* William Armbruster, "Customs defuses Miami trucker protest," *Journal of Commerce Online*, October 28, 2003, www.joc.com/.

27 *Each day, the bridge:* Daily car and truck figures for the Ambassador Bridge are based on *Transportation in Canada 2002: Transport Canada, 2002 Annual Report*, Appendix, A83, "Table A7–13: Twenty Largest Border Crossings, Cars/Other Vehicles, 1999–2002," and "Table A7–12: Twenty Largest Border Crossings for Trucks, 1999–2002." Value figure is based on "Table A2–4: Canada's Road Trade with the United States by Busiest Border Crossing Points, 2002," using a conversion rate of .7236.

32 *The transformers, which are produced: Making the Nation Safer: The Role of Science and Technology in Countering Terrorism.* Report of the Committee on Science and Technology for Countering Terrorism (Washington, DC: The National Academies Press, 2002): 182.

32 *Because of the combination of high demand:* The ports of Los Angeles and Long Beach brought in 18.9 million metric tons or 138,177,628 barrels of oil in 2001. The rest of the state's ports brought in 1.9 metric tons or 160,543,381 barrels in 2001. (Based on "U.S. Imports by U.S. Customs District and Port—2001," MARAD, Department of Transportation. In 1999, California sources accounted for 828,000 barrels per day or 48 percent of the State's total

crude oil demand. During this same period, an average of 530,000 barrels per day (30 percent) was imported from Alaska and nearly 386,000 barrels per day (22 percent) from foreign sources. (California Energy Outlook 2000, Volume II, Transportation Energy Systems, August 2000.) Because California is not connected by pipeline to refineries in other states, crude oil, gasoline, and other fuels must be brought in by marine tankers. The state's marine oil terminals are currently running at or near capacity and are concentrated in just a few locales. The state's refining capacity operates at near-maximum levels. Inventories represent only eighteen days of supply on average and replacement supplies can take up to eight weeks to reach marine terminals. According to one recent state of California report: "An upset in the petroleum system can immediately translate into tight supplies and higher prices at the pump." (2003) Integrated Energy Policy Report, Docket #02-IEP-01, Publication Number: 100–03–019F, Adopted by Energy Commission: November 12, 2003, and California Energy Commission, *California Marine Petroleum Infrastructure*, Consultant Report, California Energy Commission, Sacramento, CA, P600–03–008.

33 *They estimate the loss of production:* Stephen E. Flynn, "Beyond Border Control," *Foreign Affairs*, LXXIX:6 (Nov.–Dec., 2000): 59–60.

33 *Emergency responders in Los Angeles:* The lack of necessary protective equipment among first responder communities was highlighted most notably by the Council on

Foreign Relations Task Force Report, *Emergency Responders: Drastically Underfunded, Dangerously Unprepared* (New York: Council on Foreign Relations, July 2003). Also see chapter six in this volume for a fuller discussion.

34 *The Secretary of Transportation:* The West Coast port strike in the fall of 2002 illustrated this phenomenon well. The 11-day lockout left ships stranded at anchor and docks piled up with cargo from Los Angeles to Taipei. The total cost to the U.S. economy from that incident is estimated at up to nineteen billion dollars. See Evelyn Iritani and Marla Dickerson, "The Port Settlement; Tallying Port Dispute's Costs; Many companies managed to escape major losses, but some negative effects could be permanent," *Los Angeles Times*, November 25, 2002.

34 *A call from the governor:* Phone interview with Major Charles Anthony, a Public Affairs Officer, Department of Defense, State of Hawaii, August 18, 2003.

35 *Legend has it that after his forces:* General Charles C. Krulak, Commandant, U.S. Marine Corps, Remarks at the Council on Foreign Relations, November 17, 1997.

3 The Phony War

Page

37 *The next eight months:* In *Sartre: A Life*, Annie Cohen-Solal writes "The Phony War of 1939–40 is no war at all. From their fathers, grandfathers, and uncles, from the old soldiers, from all the veterans of World War I, they had

come to expect trenches, visible enemies, real confrontations, heroic soldiers. For over nine months they will just wait, interminably and absurdly, among other French soldiers, waiting for the Germans to attack." Annie Cohen-Solal, *Sartre: A Life*, (New York: Pantheon, 1987): 134; Sartre himself writes in his notebooks, "All at once I found myself part of a mass of men in which I had been given an exact and stupid part to play, a part I was playing . . . opposite other men, dressed as I was in military clothes, whose part was to thwart what we were doing and finally to attack." See Jean-Paul Sartre, *War Diaries: Notebooks from a Phoney War: November 1939—March 1940*, (London: Verso, 1984): 387–388.

38 *The British Expeditionary forces*: John Keegan, *The Second World War* (New York: Penguin, 1990); and John Keegan, *A History of Warfare* (New York: Vintage, 1994).

39 *The Posse Comitatus Act:* After the Civil War, the Army was used to help maintain civil order and enforce new reconstruction-era policies. Congress soon became concerned that the Army's role was becoming too politicized and subsequently enacted the Posse Comitatus Act to remove the Army from civilian law enforcement and to return it to its more traditional role. Today, the operative restriction can be found in U.S. Code—Title 18, Part 1, Chapter 67, Section 1385—use of Army and Air Force as Posse Comitatus, "Whoever, except in cases and under circumstances expressly authorized by the Constitution or Act of Congress, willfully uses any part of the Army or the Air Force as a Posse Comitatus or otherwise to exe-

cute the laws shall be fined under this title or imprisoned not more than two years, or both." See the Legal Information Institute, Cornell Law School, www4.law.cornell.edu/uscode/18/1385.html.

42 *the top Navy brass balked:* William P. Nash, Jr., "America's Commercial Seaports: An Achilles Heel, A Case Study of the Ports of Los Angeles and Long Beach" (unpublished paper, June 18, 2001).

42 *The Coast Guard is charged with protecting: Oceans and Coastal Resources, a Briefing Book,* Congressional Research Service Report 97–588-ENR, (Washington, D.C.: USGPO, May 30, 1997).

42 *with a force about the same size:* As of July 2003, the Coast Guard had 39,000 active duty personnel. U.S. Coast Guard Factfile, www.uscg.mil/hq/g-cp/comrel/factfile/index.htm; the NYPD has approximately 39,110 active duty police officers, New York Police Department, Frequently Asked Questions, www.ci.nyc.ny.us/html/nypd/html/misc/pdfaq2.html#41.

42 *deployed on a fleet of vessels:* Michael Kilian, "Coast Guard being spread too thin, report finds," *Chicago Tribune*, February 15, 2002. Of the service's fleet of cutters and other major vessels, two crafts were commissioned in 1936, and several date back to World War II.

42 *And while the Coast Guard was handed more:* Former Coast Guard Commandant Admiral James Loy made this comment on the Center for the Business of Government Radio Show on May 1, 2000: "Our service is about the same size in

terms of people as it was in the mid-1960s. And in the meantime, we have had an enormous array of challenges rolled in our direction, some from the Congress, some from various administrations over that 30 years, and some from the American people themselves, who often have an opportunity to stipulate quite clearly what it is that they're interested in some organization doing." www.businessofgovernment.org/-main/interviews/profile/index.asp?PID=14.

42 *The number of customs inspectors:* In 1992, the Customs Service had 6,239 inspectors at our ports, airports, and land border crossings. In 1998, there were 7,448 inspectors and 19,461 total employees. Now Customs and Border Protection has just 8,700 full-time customs inspectors and 328 part-time inspectors. Trade, as measured in merchandise entries, has ballooned from less than 9 million in the early 1990s to 26.1 million entries in fiscal year 2003. *Sources on personnel levels: Occupations of Department of Homeland Security Staff,* March 2003, Department of Homeland Security: The First Month, TRAC Reports, trac.syr.edu/tracreports/tracdhs/030825/occupate.html; sources on trade volume: U.S. Customs Annual Reports and press releases.

43 *Budget requests to replace ancient systems:* The Customs Modernization Act of 1993 promised the trade community a major-systems upgrade known as ACE, or Automated Commercial Environment. That program got its first down payment in fiscal year 2001 and is not due to be completed until 2005. The third stage of the program's release was delayed in late 2003 to the end of 2004. See

Paul Spillinger, "ACE receives first major appropriation," *Journal of Commerce*, July 24, 2000; and R.G. Edmonson, "Customs delayed on new ACE release," *Journal of Commerce*, September 23, 2003.

43 *Many inspection facilities still rely:* Before Sept. 11, half the northern border's 126 official crossings were unguarded at night. Typically, orange cones were simply put up in the center of the roadway to signal the border crossing was closed. See Michael Grunwald, "Economic Crossroads on the Line; Security Fears Have U.S. and Canada Rethinking Life at 49th Parallel," *Washington Post*, December 26, 2001; Senator Byron Dorgan (D-ND) caustically summarized the situation in a hearing three weeks after 9/11: "I have also seen the little orange cones which secure a good number of the northern Ports of Entry at night. All terrorists and smugglers know that the only thing that precludes them from entering this country in many locations at the Northern Border is a rubber cone. The nice ones will put the cones back after driving through," Senator Byron L. Dorgan, *Statement of Senator Byron L. Dorgan before the U.S. Senate Committee on Appropriations*, 107 Cong., 1st sess., October 3, 2001.

44 *According to a study released in July 2003:* The Partnership for Public Service, *Homeland Insecurity: Building the Expertise to Defend America from Bioterrorism*, (Washington: 2003): 5.

44 *The insular nature of the agency:* "[Federal] agencies are still not communicating with each other. . . . There have been decades of game play. Has there been progress? Yes.

But agencies are still not talking," Senator Richard Durbin (D-IL) quoted in Carol Marin, "FBI's tardy computer upgrade worth fretting about," *Chicago Tribune*, December 17, 2003.

44 *A much needed "virtual case file":* Cam Simpson, "FBI hits glitches as it joins digital age; Upgrade ordered after 9/11 attacks," *Chicago Tribune*, December 10, 2003.

45 *Before it was broken up and reassigned:* The problems the INS faced, and continues to face, were well documented in a series of scathing reports by the Government Accounting Office and Department of Justice Inspector General. See U.S. Government Accounting Office, *INS Management: Follow-up on Selected Problems*, GAO/GGD-97–132 (July 22, 1997); U.S. Government Accounting Office, *Immigration and Naturalization Service: Overview of Management and Program Challenges*, GAO/T-GGD-99–148 (July 29,1999); and "Management Challenges in the Department of Justice," Office of the Inspector General, December 1, 2001.

45 *A University of Texas study:* The Population Research Center, *An Estimate of the Number of Border Patrol Personnel Necessary to Control the Southwest Border* (Austin, TX: University of Texas at Austin, 1998).

46 *In 1997, the congressionally mandated:* "We can assume that our enemies and future adversaries have learned from the Gulf War. They are unlikely to confront us conventionally with mass armor formations, air superiority forces, and deep-water naval fleets of their own, all areas of overwhelming

U.S. strength today. Instead, they may find new ways to attack our interests, our forces, and our citizens. They will look for ways to match their strengths against our weaknesses. They will actively seek existing and new arenas in which to exploit our perceived vulnerabilities. Moreover, they will seek to combine these unconventional approaches in a synergistic way." See report of the National Defense Panel, *Transforming Defense: National Security Strategy in the 21st Century*, (Washington, D.C.: December 1997): 11, www.dtic.mil/ndp/FullDoc2.pdf.

46 *That same year, the president's Commission:* "We are convinced that our vulnerabilities are increasing steadily, that the means to exploit those weaknesses are readily available and that the costs associated with an effective attack continue to drop. What is more, the investments required to improve the situation—now still relatively modest—will rise if we procrastinate. . . . We should attend to our critical foundations before we are confronted with a crisis, not after. Waiting for disaster would prove as expensive as it would be irresponsible." See the Report of the President's Commission on Critical Infrastructure Protection, *Critical Foundations: Protecting America's Infrastructure* (Washington, D.C.: October 1997): x.

46 *In 2000, the Gilmore Commission:* "One of our fundamental assumptions has been that no single jurisdiction can handle a major terrorist attack." See the Second Annual Report to the President and the Congress of the Advisory Panel to Assess Domestic Response Capabilities For Ter-

rorism Involving Weapons of Mass Destruction, II. *Toward a National Strategy for Combating Terrorism* (Washington, D.C.: RAND, December, 2000): 29.

46 *That same year, the Interagency:* "The level of security does not even meet the Commission's definition of minimum standards or guidelines. Some ports have been working diligently to improve security, but most have not." *Report of the Interagency Commission on Crime and Security in U.S. Seaports* (Washington, D.C.: Fall 2000): 77.

46 *The primary finding was that the most likely threat:* "The combination of unconventional weapons proliferation with the persistence of international terrorism will end the relative invulnerability of the U.S. homeland to catastrophic attack. A direct attack against American citizens *on American soil* is likely over the next quarter century," *See* The U.S. Commission on National Security/21st Century, *Road Map for National Security: Imperative for Change*, the Phase III Report of the U.S. Commission on National Security/21st Century, (Washington, D.C.: February 15, 2001): viii.

46 *A major exercise held that year:* TOPOFF was a $3-million drill that tested the readiness of top government officials to respond to terrorist attacks, directed at multiple geographic locations. It was the largest exercise of its kind to date. The exercise, which took place in May 2000 in three cities in the United States, simulated a chemical weapons event in Portsmouth, New Hampshire, a radiological event in the greater Washington, D.C., area, and a bioweapons event in Denver, Colorado. The bioterror-

ism component of the exercise centered on the release of an aerosol of *Yersinia pestis*, the bacteria that causes plague. Denver was selected in part because it had received domestic preparedness training and equipment. The logistics of the distribution of antibiotics from the National Pharmaceutical Stockpile (NPS) was another process that raised concern. Local antibiotic supplies in Denver had been depleted early in the crisis. Material from the NPS was requested quickly during the outbreak, and its delivery was approved by the Surgeon General and the director of the Centers for Disease Control. The delivery of components of the NPS to Denver was largely notional, but material that resembled components of the stockpile was flown to an airport in Denver. Stockpile material delivered to the airport for the purposes of the drill was at one point being unbundled by a single individual who "had to count individual pills and put them into plastic baggies." Before she could even begin, there was a six-hour delay during which terrible (notional) traffic was negotiated "in order to get the plastic baggies from Safeway." In addition to deciding to deliver antibiotics to hospitals, the governor's Emergency Epidemic Response Committee decided to distribute prophylactic antibiotics through central antibiotic distribution facilities that were termed "points of distribution" (PODs). Although the governor's Emergency Epidemic Response Committee decided to open multiple PODs, only a sample POD was exercised during TOPOFF. One observer remarked, "They could take care of only 140 people per hour at the mass antibiotic prophylaxis [POD] center. For a city of one million, that's pitiful." Other concerns regarding the POD included the following: "No written guidelines were given out with the dispensed antibiotics. . . . Directions were too difficult to understand," and, "Persons were allowed to take their antibiotics from a box, with little oversight regarding how much each took." Another commented, "What we needed from CDC were people to help us treat and give out antibiotics, not epidemiologists." As part of the exercise, fifty trainees from the Bureau of Alcohol, Tobacco and Firearms had been instructed to cause as much unrest as possible at the POD. The unrest was largely notional, but one observer concluded that in a real epidemic, each POD would "require several hundred people to staff it and provide security." No such staffing plan existed for the POD. Despite this, one of the major hospitals had to drop out of the exercise prematurely because it had so many actual patients that needed treatment that it could not spare the resources to participate even in the notional "on-paper" elements of the exercise. A number of other serious problems were cataloged. "There were not enough places to put sick people, triage people, put dead bodies." Hospitals were competing for ventilators. It was not clear which health-care workers should be wearing personal protective equipment or what form of protection was appropriate. See Donald A. Henderson, Thomas V. Inglesby Jr., and Tara O'Toole, "A Plague on Your City: Observations from TOPOFF," *Clinical Infectious Diseases*, 32 (2001): 436–45.

47 *Another bio-threat exercise sponsored:* The war game involved senior-level officials, including then-governor

Frank Keating of Oklahoma, former presidential adviser David Gergen, and former director of Central Intelligence James Woolsey. Among the findings: (1) an attack on the United States with biological weapons could cause massive civilian casualties, breakdowns in essential institutions, disruption of democratic processes, civil disorder, loss of confidence in government, and reduced U.S. strategic flexibility; (2) the government currently lacks adequate strategies, plans, and information systems to manage a crisis of this type or magnitude; (3) public health is now a major national security issue; (4) constructive media relationships become critical for all levels of government; (5) containing the spread of a contagious disease delivered as a bioweapon will present significant ethical, political, cultural, operational, and legal challenges. See John Hamre and Sam Nunn, "Testimony before the House Subcommittee on National Security, Veterans Affairs, and International Relations," July 23, 2001.

48 *America remains dangerously unprepared: America: Still Unprepared, Still in Danger,* Report of an Independent Task Force (New York: Council on Foreign Relations, October 2002).

48 *For instance, another Council on Foreign Relations Task Force Report: Emergency Responders: Drastically Underfunded, Dangerously Unprepared,* Report of an Independent Task Force (New York: Council on Foreign Relations, July 2003).

49 *Things have certainly improved:* The Aviation and Transportation Security Act was signed into law on November

19, 2001 by President Bush. It requires that all cargo aboard commercial flights be screened. Despite this requirement, a Congressional Research Service study in November 2003 reported that less than 5 percent of cargo placed on passenger planes is being screened. See Congressional Research Service, *Air Cargo Security* (Washington, D.C.: U.S.GPO, September 11, 2003); and U.S. General Accounting Office, *Testimony before the Committee on Commerce, Science and Transportation, U.S. Senate, Aviation Security: Efforts to Measure Effectiveness and Address Challenges*, GAO-04–232T (Washington, D.C.: November 5, 2003), www.gao.gov/new.items/d04232t.pdf.

49 *As long as air cargo comes* : U.S. General Accounting Office, *Testimony before the Committee on Government Reform, U.S. House of Representatives: Aviation Security: Efforts to Measure Effectiveness and Strengthen Security Program*, GAO-04–285T, 108 Cong., 1st sess. (Washington, D.C.: November 20, 2003), reform.house.gov/UploadedFiles/GAO%20%20-Berrick%20Testimony.pdf.

50 *Only in November 2003:* See "Analyst rips new air cargo security rules," *Journal of Commerce Online*, November 24, 2003, www.joc.com/; and R.G. Edmonson, "TSA imposes new security measures," *Journal of Commerce Online*, November 18, 2003, www.joc.com/.

51 *The president's charter for the office:* "Executive Order Establishing the Office of Homeland Security and the Homeland Security Council," October 8, 2002, www.whitehouse.gov/news/releases/2001/10/20011008–2.html.

52 *As Governor Ed Rendell:* Robert Pear, "Governors Get Sympathy from Bush but No More Money," *The New York Times*, February 25, 2002.

53 *First, it fails to take into account:* Ray Scheppach, executive director of the National Governors' Association, described the 2002 states' budget outlook as the bleakest since World War II and the budget situation in 2003 was even bleaker. He's also quoted as saying: "It's clearly the worst since we've been keeping statistics." See Jodi Wilgoren, *New York Times*, June 27, 2003. According to the National Governors Association, National Association of State Budget Officers, *The Fiscal Survey of States* (Washington, D.C.: June 2003) "37 states cut their budgets by $14.5 billion, the highest dollar amount of cuts in the history of the Fiscal Survey. . . . By comparison, during the last recession in fiscal 1992, 35 states reduced their budgets by nearly $4.5 billion." On the limited homeland-security progress in Pennsylvania, Texas, Washington, and Wisconsin, see Donald F. Kettl, *The States and Homeland Security: Building the Missing Link*, The Century Foundation's Homeland Security Project Working Group on Federalism (New York: The Century Foundation, 2003), www.tcf.org/Publications/HomelandSecurity/kettl.pdf. According to the State of California's *Governors Budget, May Revision 2004*, California's adjusted budget gap was $38.2 billion in 2003. In response to a growing budget gap, Oregon officials proposed Ballot Measure 28, giving Oregonians the choice between raising taxes and slashing state programs. Measure 28 was defeated and Oregon voters voted to trim vital health services as well as public school budgets. Many schools were forced to close doors before the end of

the school year: "The failure of the proposal means cuts in the state trooper force, assistance for low-income senior citizens and the disabled, community mental health efforts and school funding, the *Statesman Journal* newspaper reports" See "Oregon voters reject proposed tax hike," United Press International, January 29, 2003.

55 *"Ruin is the destination toward which all men rush:* "The Tragedy of the Commons," Garrett Hardin, *Science* 162 (December 13, 1968): 1243–48.

4 Security Maturity

Page

62 *Cars can be dangerous:* The Bureau of Transportation Statistics, National Transportation Statistics 2003, "Table 2–1: Transportation Fatalities by Mode," www.bts.gov/.

65 *In 2002, three million trucks:* According to Bureau of Transportation Statistics, 1,441,653 trucks came into the United States at Laredo in 2002; approximately as many entered Mexico at the border crossing in Nuevo Laredo. In 1994, 668,000 incoming trucks were recorded. Sources: "Incoming Truck Crossings, U.S.-Mexican Border, 2002," U.S. Department of Transportation, Bureau of Transportation Statistics, www.bts.gov/, based on data from U.S. Customs Service, Mission Support Services, Office of Field Operations, Operations Management Database; and "Table 25: Top 20 NAFTA Border Truck Crossings Into the United States, 1994, 2000, 2001," *U.S. International Trade and Freight Transportation Trends*, U.S. Department of Trans-

portation, Transportation Statistics, 2003, www.bts.gov/publications/us_international_trade_and_freight_transportation_trends/2003/.

65 *The annual job turnover rate:* Interviews with U.S. Customs and Immigration officials during a field visit to Laredo, Texas August 20–21, 2001.

66 *That help is provided by criminal networks:* Peter Andreas, *Border Games: Policing the U.S.—Mexico Divide* (Ithaca: Cornell University Press, 2000).

70 *What they have achieved provides a template:* This section is based on on-site visits and interviews at Logan Airport, Boston, MA, hosted by George Naccara, the Transportation Security Administration's Director of Security on February 10, 2003 and October 14, 2003.

5 What's in the Box?

Page

81 *Now, as head of Hutchison Port Holdings:* See Hutchison Port Holdings Web site: www.hph.com.hk/corporate/introduction.htm.

81 *While Hutchison owns no terminals:* Interview with Garry Gilbert, corporate adviser, North America, Hutchison Port Holdings, Washington, D.C., January 21, 2004.

82 *Today, the two largest container ports:* Container movements are counted in the rather anachronistic measurement of "twenty-foot equivalent units" or TEUs. However, about two-thirds are actually forty feet long, similar

in size to a large moving truck or freight car. In 2002, the Port of Hong Kong handled 19,144,000 TEUs and the Port of Singapore handled 16,940,900 TEUs for a total of 36,084,900 for the year or about 3 million a month as measured in TEUs. For Port of Singapore statistics see: Maritime and Port Authority of Singapore "Total Container Throughput (in 2000 TEUs)," www.mpa.gov.sg/homepage/portstats.html; for Port of Hong Kong statistics see: Hong Kong Special Administrative Region of The People's Republic of China, Government Information Centre, "Inward/Outward Container Throughput by Laden/Empty Containers," www.mardep.gov.hk/en/publication/pdf/portstat_1_y_b1.pdf.

84 *Because 90 percent of the world's:* "When trade and security clash," Special Report, *The Economist*, April 4, 2002.

86 *Part of the challenge derives from:* The butterfly effect is a central idea in the branch of mathematics known as Chaos Theory. James Gleick, the former science writer for *The New York Times*, provides an excellent description of the early years of this field in his book *Chaos* (New York: Penguin, 1998).

86 *The chain of events that led to 263 power plants:* U.S.–Canada Power System Outage Task Force, *Interim Report: Causes of the August 14th Blackout in the United States and Canada* (November 2003).

87 *In the aftermath of the 9/11 attacks: The Port Protection Act of 2002*, H.R. 5420, 107th Congress, September 19, 2002.

87 *Nadler's solution, which he incorporated into a bill:* H.R. 1010, "The Port Protection Act of 2003," 108th Congress, February 27, 2003, reads in part: "Sec. 15102. Inspection of cargo vessels prior to entry into United States port: (a) The Commandant of the Coast Guard shall board and inspect each vessel carrying cargo destined for the United States, at least 200 miles from the United States. (b) An inspection under this section shall include physical inspection to verify that—(1) the cargo containers on the vessel have not been tampered with; and (2) the remainder of the vessel, including but not limited to noncontainerized cargo, the engine room, living quarters, bathrooms, and hull, does not contain a chemical, biological, or nuclear weapon."

89 *These containers have up to thirty days:* The Customs Service's own Performance and Annual Report for Fiscal Year 2002 found that "Customs did not adequately monitor the movement of in-bond merchandise and did not measure the extent of compliance with Customs policies in 2002." The report went on to describe further problems: "In the ports subject to examination, we noted that the Inspectors do not always verify the quantity and type of cargo prior to the in-bond movement. In some situations the reports of movements did not match the arrived-at-destination reports. System and procedural limitations can complicate or even prevent adequate reconciliation and matching of movements. . . . Reports of overdue in-bond shipments showed that a number of in-bond movements were not always resolved in a timely manner . . ." See U.S. Customs Service, *Report on Internal*

Control over Financial Reporting, Performance and Annual Report for Fiscal Year (Washington, D.C.: 2002): 118.

89 *On average, an overseas container will pass:* Interview with Garry Gilbert, Corporate Adviser, North America, Hutchison Port Holdings, Washington, D.C., January 21, 2004.

90 *Based on a computerized Automated Targeting System:* "With respect to the sources and types of information, ATS [Automated Targeting System] relies on the manifest as its principal data input, and CBP [Customs and Border Protection] does not mandate the transmission of additional types of information before a container's risk level is assigned. Terrorism experts, members of the international trade community, and CBP inspectors at the ports we visited characterized the ship's manifest as one of the least reliable or useful types of information for targeting purposes. In this regard, one expert cautioned that even if ATS were an otherwise competent targeting model, there is no compensating for poor input data. Accordingly, if the input data are poor, the outputs [i.e., the risk assessed targets] are not likely to be of high quality. Another problem with manifests is that shippers can revise them up to 60 days after the arrival of the cargo container. According to CBP officials, about one third of these manifest revisions resulted in higher risk scores by ATS—but by the time these revisions were received, it is possible that the cargo container may have left the port. These problems with manifest data increase the potential value of additional types of information." U.S. Government Accounting Office, Statement of Richard M.

Stana, Director, Homeland Security and Justice, *Homeland Security: Preliminary Observations on Efforts to Target Security Inspections of Cargo Containers, Testimony Before the Subcommittee on Oversight and Investigations, Committee on Energy and Commerce, U.S. House of Representatives*, GAO-04–325T, (Washington, D.C.: December 16, 2003): 11; Rob Quartel, a former federal maritime commissioner, and now CEO of FreightDesk Technologies, recently made the case that the manifest is a fairly poor means of garnering critical information to detect a terrorist threat in his *Journal of Commerce* piece, "What's in the box? It doesn't matter," November 3, 2003. Quartel writes, "Where terrorism is concerned, however, it's not the contents but the context. People, processes and events, rather than goods, are what are relevant. The 'virtual border' and the derivative 24-hour rule were animated by this idea—that everyday commercial transactions, fused with intelligence data, can provide a picture of the confluence of events and parties in which trace patterns of terrorist activity can be seen. Who touched a shipment, where it's been or is going, who paid for or insured it, who owned the ship or truck that moved it, perhaps C-TPAT participation, are together more likely indicators of risk than the bill of lading. No one needs know what's in—or what someone says is in—a container to conclude that there's a need for a physical inspection."

96 *The top five maritime loading centers:* The combined port totals in calendar year 2002 for the ports of LA–Long Beach, Seattle–Tacoma, New York, Charleston, and

Oakland equals 6,814,000 TEUs. The remaining twenty-four major container operations handled 2,851,000 TEUs. The top ten handle approximately 98 percent. Statistics taken from Maritime Administration, Department of Transportation, "U.S. Waterborne Foreign Trade, Containerized Cargo, Top 30 U.S. Ports, Calendar Year 2002 (Based on data from the Port Import/Export Reporting Service, PIERS), www.marad.dot.gov/Marad_Statistics/Con-Pts-02.htm.

96 *Fifty percent of the containers:* Imports vs. exports of loaded boxes work out to about a three-to-one ratio. The ports of Los Angeles and Long Beach shipped about three million empty boxes back to Asia last year. See Bill Mongelluzzo, "Record year for Los Angeles–Long Beach," *Journal of Commerce Online*, January 21, 2004, www.joc.com/.

96 *Right now the odds stand at about 10 percent:* The 10-percent probability of detection was arrived at using a mathematical model which examined a eleven-layer security system consisting of shipper certification, container seals, and a targeting software system, followed by passive (neutron and gamma), active (gamma radiography) and manual testing at overseas ports. See Lawrence M. Wein, Alex H. Wilkins, Manas Baveja, and Stephen E. Flynn, "Preventing the Importation of Illicit Nuclear Materials in Shipping Containers," Unpublished Technical Paper, Stanford University (October 2003).

99 *William Hamlin, the man responsible:* William Hamlin, Keynote Address (Intermodal Freight Technology Working Group, Minneapolis, Minnesota, May 18, 2003).

101 *These units cost about one million dollars each:* Science Applications International Corporation (SAIC), L-3 Communications, and AS&E, make these systems.

105 *"As with any new proposal":* Robert C. Bonner, U.S. Customs Commissioner, speech before the Center for Strategic and International Studies (CSIS), Washington D.C., January 17, 2002: 21.

107 *However, once the enforcement deadline:* "Frequently Asked Questions: 24-Hour Vessel Manifest Rule," U.S. Bureau of Customs and Border Protection, December 12, 2003, www.cbp.gov/xp/cgov/import/carriers/24hour_rule/.

107 *In addition, by the summer of 2003:* "Ports in CSI," Office of International Affairs, U.S. Customs & Border Protection, www.cbp.gov.

108 *This small demonstration project:* Testimony of Peter Hall, U.S. Attorney, District of Vermont, *U.S. Senate Governmental Affairs Committee hearing on "Cargo Containers: The Next Terrorist Target?"* 108 Cong., 1st sess. (Washington, D.C., March 20, 2003).

109 *Based on preliminary economic analysis:* (Network Visibility: Leveraging Security and Efficiency in Today's Global Supply Chains) Phase One Review, *A Smart & Secure Tradelanes White Paper* (November, 2003).

109 *Also, all civilized nations will:* Michael Sheridan, James Clark, and Edin Hamzic, "The snakehead trail," *Sunday Times* (London), June 25, 2000. Another tragic incident

involving death of migrants smuggled in a container took place in Waterford in Southern Ireland on December 7, 2001. The bodies of eight dead refugees and five survivors were found in a container full of office furniture being delivered to a business park. The container was shipped from the Belgian port of Zeebrugge to Waterford, a two-day voyage, after a journey by road through Italy and Germany. Nine of the thirteen refugees came from Turkey, mostly from two families, as well as from Algeria and Albania. The five deaths included two boys, ages four and nine, a ten-year-old girl, a teenage male, three male adults, and an adult woman. See "Survivors of freight container tragedy can stay in Ireland, government says," Associated Press, December 10, 2001.

110 *Since it is hard to figure out what is in the box*: Interview with Michael Surridge, Commissioner of Revenue Protection, Government of Jamaica, in Kingston, Jamaica, on August 16, 2000.

6 Protect and Respond

Page

111 *U.S. intelligence officials had initially believed:* John Solomon, "9/11 Planner Tells of Plot's Origins," *Washington Post*, September 22, 2003.

113 *Finally, they can target market-ready foods:* Henry S. Parker, *Agricultural Bioterrorism: A federal strategy to meet the threat*, McNair Paper 65, Institute for National Strategic Studies,

National Defense University, Washington, D.C., (March 2002), www.ndu.edu/inss/McNair/mcnair63/mcnair63.pdf.

113 *Food production makes up nearly 10 percent:* Statement of Senator Susan M. Collins, *Hearing on Agroterrorism: The Threat to America's Breadbasket, U.S. Senate Government Affairs Committee*, 108th Cong., 1st sess., November 19, 2003, www.senate.gov/~govt-aff/_files/111903chalk.pdf.

113 *It is also an important component:* Dr. Peter Chalk, *Hitting America's Soft Underbelly: The Potential Threat of Deliberate Biological Attacks Against the U.S. Agricultural and Food Industry*, prepared for the Office of the Secretary of Defense, Rand Corporation, (2004): ix.

114 *One California study estimated:* Testimony of Dr. Peter Chalk, *Hearing on Agroterrorism: The Threat to America's Breadbasket, Government Affairs Committee of the U.S. Senate*, 108th Cong., 1 sess., November 19, 2003, www.senate.gov/~govt-aff/_files/111903chalk.pdf.

114 *at which point most of America's:* The drill was funded by the USDA and run by the Central States Animal Emergency Coordinating Council. See Mark Murray et al., "Hardening the Targets," *National Journal*, August 10, 2002.

114 *In doing so, they closed their markets:* U.S. beef-export revenue for 2002 was 3.2 billion according to the U.S. Beef Export Federation web site, usmef.org/Statistics2002/export02_12_BeefPlus.pdf.

114 *By way of illustration, it would take a ditch:* Telephone

interview with Paul Engler, Chairman, Cactus Feeders, Amarillo, Texas, on January 23, 2004.

114 *A major outbreak could easily involve:* The Trust for America's Health issued a report in August 2003 that estimates more than two hundred government offices and programs would be involved in responding to a major food-disease outbreak. See "Animal-Borne Epidemics Out of Control: Threatening the Nation's Health," Issue Report, Trust for America's Health, August 2003: 16, www.healthyamericans.org/reports/files.animalreport.pdf.

115 *The group brings together beef producers:* Phyllis Jacobs Griekspoor, "VeriPrime certifies Abbott Laboratories for fast mad-cow test," *Wichita Eagle*, January 21, 2004.

116 *These teams include animal control:* Corine Hegland, "What if onions were a terrorist plot," *National Journal*, January 10, 2003.

117 *A month and a half after:* Shankar Vedantam, "U.S. Ends Investigation of Mad Cow Case; Officials Fail to Find Two-Thirds of Animals At Risk of Infection," *Washington Post*, February 10, 2004.

118 *For instance, public water utilities:* U.S. General Accounting Office, *Homeland Security: Voluntary Initiatives Are Under Way at Chemical Facilities, but the Extent of Security Preparedness Is Unknown*, GAO-03–439 (Washington, D.C.: March 2003): 1, www.gao.gov/highlights/d03439high.pdf.

118 *Detonating a tank of chlorine gas:* Claudia Copeland and Betsy Cody, Resource, Science and Industry Division,

Terrorism and Security Issues Facing the Water Infrastructure Sector, Congressional Research Service Report for Congress (Washington, D.C.: USGPO, May 21, 2003), www.ncseonline.org/nle/crsreports/03Jun/RS21026.pdf; and Carl Prine from the *Pittsburgh Tribune Review* wrote a number of *Tribune* articles in which he visited several sites, including water treatment facilities, that use toxic chemicals. Prine reported: "At the Wilkinsburg Penn Joint Water Treatment Facility in Verona, a broken fence and an unlocked door allowed a Trib(une) reporter to reach 20 tons of chlorine gas and millions of gallons of drinking water. If the chlorine tank ruptured, the gas could lap neighborhoods up to three miles away, threatening more than 100,000 people." See "Chemical Sites Still Vulnerable," *Pittsburgh Tribune Review*, November 16, 2003.

118 *According to the Environmental Protection Agency:* U.S. General Accounting Office, *Homeland Security: Voluntary Initiatives Are Under Way at Chemical Facilities, but the Extent of Security Preparedness Is Unknown*, GAO-03–439 (March 2003): 4, www.gao.gov/highlights/d03439high.pdf.

119 *After 9/11, Senator Jon Corzine:* On October 31, 2001, Senator Corzine introduced: *Chemical Security Act of 2001*, S. 1602, 107th Cong., 1st sess., October 31, 2001.

119 *The chemical industry rallied nearly:* "EPA Drops Chemical Security Effort; Agency Lacks Power to Impose Anti-Terror Standards, Lawyers Decide," *Washington Post*, October 3, 2002.

119 *As a result, a more industry-friendly:* On May 5, 2003, Senator James Inhofe introduced: *Chemical facilities Security Act of 2003*, S.994, 1st 108th Cong., 1st sess., May 5, 2003.

120 *Basic measures such as posting:* U.S. General Accounting Office, *Homeland Security: Voluntary Initiatives Are Under Way at Chemical Facilities, but the Extent of Security Preparedness Is Unknown*, GAO-03–439 (March 2003): 12, www.gao.gov/highlights/d03439high.pdf.

121 *A study prepared by Michael Levi:* Federation of American Scientists Web site, "Testimony of Dr. Henry Kelly, President of the Federation of American Scientists before the Senate Committee on Foreign Relations, March 6, 2002," www.fas.org/ssp/docs/030602-kellytestimony.htm.

122 *But the budget for this disposal:* According to Dr. Henry Kelly, President of the Federation of American Scientists, funding for the DOE OFF-Site Recovery Project (OSRP) "has been repeatedly cut in the FY2001 and 2002 budgets and the presidential FY2003 budget proposal, significantly delaying the recovery process. In the cases of FY01 and FY02, the 25 percent and 35 percent cuts were justified as money being transferred to higher priorities; the FY03 would cut funding by an additional 26 percent." See Federation of American Scientists Web site, "Testimony of Dr. Henry Kelly, President of the Federation of American Scientists before the Senate Committee on Foreign Relations, March 6, 2002," www.fas.org/ssp/docs/030602-kellytestimony.htm.

123 *The vials were notionally:* John Heilpern, "Inspectors find

lax security for potential bioterrorism agents at federal research labs," Associated Press, November 20, 2003.

124 *The potential dawning:* Jeronimo Cello, Aniko V. Paul, Eckard Wimmer, "Chemical Synthesis of Poliovirus DNA: Generation of Infectious Virus in the Absence of Natural Template," *Science* 297 (August 9, 2002): 1016–18.

125 *The federal government:* The *8-City Enhanced Terrorism Surveillance Project* is a CDC initiative. For more information, visit the Web site Centers for Disease Control's Epidemiology Program Office, Division of Public Health Surveillance and Informatics: www.cdc.gov/epo/dphsi/8city.htm.

125 *Finally, the Strategic National Stockpile:* Effective on 125 1, 2003, the National Pharmaceutical Stockpile became the Strategic National Stockpile (SNS). The SNS has twelve-hour "Push Packages" containing pharmaceuticals, antidotes, and medical supplies designed to meet the demands of a bioterror event. Under President Bush's budget proposal for FY05, the SNS was moved out of DHS and back into HHS. For more information visit the CDC Web site: www.bt.cdc.gov/stockpile/index.asp.

125 *Public health agencies have:* Testimony of Thomas L. Milne, Executive Director, National Association of County and City Health Officials, *U.S. Senate Hearing on Public Health Preparedness for Terrorism before the Committee on Governmental Affairs*, 107th Congress, 2 sess., April 18, 2002.

125 *The CDC has just begun undertaking:* John Loonsk, Associ-

ate Director for Informatics at the CDC outlined the problems that the CDC faces in getting its Public Health Information Network up and running. He made the following critical points: "Most public health software still focuses almost exclusively on the primary user's needs; the current public health information cycle (clinical event to response) is too long and frequently involves the manual exchange of data; almost all Anthrax test results [from the fall 2001 incidents] were communicated verbally over the phone; healthcare information technology's current state is fragmented and heterogeneous; most clinical care sites still do not regularly report data electronically to public health." John Loonsk, Centers for Disease Control and Prevention, "Public Health Information Network," Plenary Presentation, May 13, 2003.

125 *According to a study of emergency responses:* Edward H. Kaplan, David L. Craft, and Lawrence M. Wein, "Emergency response to a smallpox attack: The case for mass vaccination," *Proceedings of the National Academy of Sciences* 99:6 (August 6, 2002): 10935–40.

126 *These findings were confirmed in a classified:* Judith Miller, "U.S. Has New Concerns About Anthrax Readiness," *The New York Times*, December 28, 2003.

126 *Half the medical facilities:* U.S. Government Accounting Office, *Hospital Preparedness: Most Urban Hospitals Have Emergency Plans but Lack certain Capacities for Bioterrorism Response*, GAO-03–924 (August 2003), www.gao.gov/new.items/d03924.pdf.

126 *The American Hospital Association:* Janice Hopkins Tanne, "Preventing Dark Winter," *Carnegie Reporter*, I:4, Spring 2002: 12.

127 *Today, we are fighting a different kind of war:* Senate Committee on Appropriations, *Testimony of Martin O'Malley, Mayor of Baltimore, Maryland on behalf of the U.S. Conference of Mayors before the U.S. Senate Committee on Appropriations hearing on Homeland Security*, 107 Cong., 2 sess., April 10, 2002.

128 *Furthermore, with few exceptions, first-responder commanders:* The issue of interoperable communications remains a largely unresolved issue despite the fact that it was well documented that it hampered the response to virtually every major incident well before the 9/11 attacks on the World Trade Center, including, most notably, the Columbine High School shootings on May 15, 2000, and the Oklahoma City Bombing incident on April 19, 1995. Tragically, on 9/11, an estimated 121 firefighters died when the World Trade Center's north tower fell because they did not get the same information that the NYPD officers received. See Jim Dwyer, Kevin Flynn, and Ford Fessenden, "Fatal Confusion," *The New York Times*, June 7, 2002. "So poor were communications that on one side of the trade center complex, in the city's emergency management headquarters, a city engineer warned officials that the towers were at risk of 'near imminent collapse,' but those he told could not reach the highest-ranking fire chief by radio. Instead, a messenger was sent across acres, dodging flaming debris and falling bodies, to deliver this assessment in per-

son. He arrived with the news less than a minute before the first tower fell." See Jim Dwyer, "Before the Towers Fell, Fire Dept. Fought Chaos," *The New York Times*, January 30, 2002. According to the Fifth and Final Report of the Gillmore Commission, "Throughout the public safety community, the communication systems currently in place often lack the capability for diverse responders to communicate with one another at multiagency operations. While these problems are often most serious among agencies from different jurisdictions or different levels of government, even agencies from the same jurisdiction frequently cannot readily communicate due to technical problems or differences in their organizations' procedures and communications practices." *The Fifth Annual Report to the President and the Congress of the Advisory Panel to Assess Domestic Response Capabilities for Terrorism Involving Weapons of Mass Destruction* "V. Forging America's New Normalcy," RAND, December 15, 2003.

128 *In many cases, the sensors that:* "America—Still Unprepared, Still in Danger," (New York: Council on Foreign Relations, 2002).

131 *Such an approach is taken by the Coast Guard:* For a discussion of how the Coast Guard's approach to maritime safety can be adopted to the imperative of maritime security, see Joseph E. Vorbach III, *Moving beyond state-centric responses to transnational security threats*, Dissertation, Medford, MA: Fletcher School of Law and Diplomacy, 2001, and "Critiquing Traditional Responses to Cocaine and Heroin Trafficking," in *The United States and Europe:*

Policy Imperatives in a Globalizing World, ed. Howard M. Hensel, 205–229. Burlington, VT: Ashgate, 2002.

7 Mobilizing the Home Front

Page

137 *So this well-trained operative:* The story of Ressam's arrest is chronicled in a series of investigative reports. "The Terrorist Within," *Seattle Times*, June 23—July 7, 2002; and Mike Carter, "Ressam: The trial and errors exposed as the terrorism case unfolds in court," *Seattle Times*, March 25, 2001. For the details on the explosive materials that Ressam was carrying, see Testimony of Gary Stubblefield, president, GlobalOptions, *Hearing on International Terrorism and Immigration Policy, U.S. House of Representatives, Committee on Judiciary, Subcommittee on Immigration and Claims*, 106th Cong., 2 sess., January 26, 2000; and Andrew Duffy, "Video of blasts stuns jurors at Ressam trial: Simulation designed to illustrate power of explosives taken from car," *Ottawa Citizen*, March 29, 2001. Some additional details are based on interviews with U.S. and Canadian law enforcement officials.

139 *But in the spring of 2002:* Stephen Brill recounts the process in detail in his book *After: How America Confronted the September 12 Era* (New York: Simon & Schuster, 2003).

139 *Then, in a move that took Washington:* "Remarks by the President in Address to the Nation," The White House, June 6, 2002.

140 *Private-sector management consultants:* For more information on the pitfalls that accompanied the biggest private sector merger in history see: Nina Munk, *Fools Rush In: Steve Case, Jerry Levin, and the Unmaking of AOL Time Warner* (New York: HarperBusiness, 2004); and Alec Klein, *Stealing Time: Steve Case, Jerry Levin and the Collapse of AOL Time Warner* (New York: Simon & Schuster, 2003).

140 *Most observers acknowledge that:* The insular nature of the armed services was illustrated by the fact that it took the Goldwater–Nichols Act of 1986 to mandate joint interservice assignments as a condition of future promotion for senior members of the military. Prior to that, senior uniformed officers interacted with one another only on an exceptional basis, and liaison assignments were often seen as career-damaging.

142 *The Pentagon has been keen to maintain:* The Department of Defense's Northern Command (NORTHCOM) differentiates "homeland security" from "homeland defense" in this way: "Terrorism targeted against the United States is fundamentally a homeland security (HLS) matter that is usually addressed by law enforcement agencies. DoD and NORTHCOM's roles in support of HLS are limited to homeland defense and civil support. . . NORTHCOM's homeland defense mission is directed against military threats emanating from outside the United States. NORTHCOM can also provide support to lead federal agencies when directed by DoD." See www.northcom.mil/index.cfm?fuseaction=s.who_combat.

142 *In the first ten months:* Interview with the Office of Legislative Affairs, U.S. Department of Homeland Security, February 5, 2004.

143 *Back at the office, DHS officials:* Shaun Waterman, "Cox: Dual oversight hurting security," United Press International, October 16, 2003.

144 *However, in the House of Representatives:* The Committee has fifty members—twenty-seven Republicans and twenty-three Democrats—and is organized as a mirror organization to DHS. Most members are chairman or ranking members from standing committees that have previously had some jurisdiction over homeland security. There are five subcommittees: Infrastructure and Border Security; Emergency Preparedness and Response; Subcommittee on Cybersecurity; Science, and Research and Development; Subcommittee on Intelligence and Counterterrorism; and the Subcommittee on Rules. For more information, visit the Select Committee on Homeland Security web site: hsc.house.gov/.

144 *While the Select Committee has very capable:* According to Rep. Jim Turner, "This is not solely a personality issue. It's an institutional problem that Congress has," Turner points out. "Once [jurisdiction] has been configured in a given way . . . then no committee readily gives it up." Shaun Waterman, "Cox: Dual oversight hurting security" United Press International, October 16, 2003.

144 *The result is that DHS officials:* According to Rep. Christopher Cox, "the department has complained bit-

terly that they are required to come to the Hill to testify before far too many committees and subcommittees in both chambers." Shaun Waterman, "Cox: Dual oversight hurting security" United Press International, October 16, 2003.

145 *In the process, the group has discovered:* The group is known as the "Partnership to Defeat Terrorism."

145 *While it regularly meets and supports the work:* "A complete list of audits or studies under way by the GAO is available in the Board's annual *Budget Review*, which is sent to Congress during the first quarter of each calendar year. Monetary policy, which is exempt from audit by the GAO, is monitored directly by Congress through written reports, including the semi-annual Humphrey–Hawkins reports, prepared by the Board of Governors," Board of Governors of the Federal Reserve System, "The Federal Reserve System: Purposes and Functions," Washington, D.C., 1994, Chap. 1.7, www.federalreserve.gov/pf/pf.htm.

149 *They are essentially forums:* For more information on the history and organizational structure of the FBI's Joint Terrorism Task Force see: Congressional Statement, Federal Bureau of Investigation, "Statement for the Record" of Assistant Special Agent in Charge Thomas Carey, Washington Field Office, Federal Bureau of Investigation on Communication with the Law Enforcement Community before the United States Senate Committee on the Judiciary," Washington, D.C., November 13, 2001, www.fbi.gov/congress/congress01/carey111301.htm; and Congressional Statement, Federal Bureau of Investigation, "Statement for the Record

of Robert S. Mueller, III Director, Federal Bureau of Investigation on Homeland Security before the Senate Committee on Governmental Affairs," June 27, 2002, financialservices.house.gov/media/pdf/091902rm.pdf.

152 *Once the fund was financed to its authorized* : The authority in the Oil Pollution Act to replenish the fund had a sunset clause that ended in 1994 and was not renewed. The act also included authority to borrow funds immediately from the U.S. Treasury to replace funds paid out after a pollution incident, with the loan to be paid off after the fees were collected. This authority also lapsed with the 1994 sunset clause. For a comprehensive description of this unique legislation, see Lawrence I. Kiern, "Liability, Compensation, and Financial Responsibility Under the Oil Pollution Act of 1990: A Review of the First Decade," *Tulane Maritime Law Journal* 24:2 (Spring 2000): 481–592.

153 *The Federal Homeland Security System board:* The Office of Domestic Finance of the U.S. Treasury Department has established a Terrorism Risk Insurance Program authorized by the Terrorism Risk Insurance Act of 2002 to provide a way to cap liability in the event of acts of terror. However, there is no mandate that private entities purchase terrorism insurance, only that insurers make this insurance available to their clients. The Treasury Department is currently evaluating whether even these "mandatory availability" provisions should be extended for policies that are issued or renewed in 2005. See www.treasury.gov/offices/domestic-finance/financial-institution/terrorism-insurance/.

154 *It also helps to ensure that there would be an organized process:* Interview with Daniel F. Sheehan, International Security Consultant and former director National Pollution Funds Center, on February 16, 2004. The value of this kind of a framework was not lost on the authors of the Maritime Transportation Security Act of 2002. That legislation authorizes the creation of private-public Area Security Committees, but so far there has been no funding provided to put these committees in place. However, the U.S. Coast Guard has directed its local level private-public Harbor Safety Committees, which are chaired by the Coast Guard Captain of the Port, to include security in their mission.

154 *Now the annual average is four thousand barrels:* Oil Companies International Marine Forum, "The U.S. Oil Pollution Act of 1990 ('OPA 90'), Why Has It Been So Successful At Reducing Spills," www.ocimf.com/downloaddocument.cfm?documentid=943.

154 *But, like the Fed, it would be better:* In a 1997 essay in *Foreign Affairs*, Alan Blinder, the former vice chairman of the Board of Governors of the Federal Reserve system, argues there are three good reasons the Fed was organized as an independent institution. First, monetary policy is a technical subject that specialists are better equipped to handle than amateurs. Second, the effects of monetary policy take a long time to work and so require patience and a focus on the long term. Lastly, the pain of fighting inflation comes well in advance of the benefits. Since politicians are focused on the near term and avoiding

anything that might translate into an unhappy electorate, they find it impractical to exercise the discipline it takes to be a central banker. Blinder maintains that "the argument for the Fed's independence applies just as forcefully to many other areas of governmental policy" such as health policy, tax policy, and environmental policy. See Alan S. Binder, "Is Government Too Political?" *Foreign Affairs* 76:6 (November/December 1997): 115–126. Fareed Zakaria also embraces the Fed model as a way to insulate policies that require an expert, long view approach from politics. See his concluding chapter in *The Future of Freedom: Illiberal Democracy at Home and Abroad* (New York: W. W. Norton, 2003).

8 Fear, Paranoia, and the Gathering Storm

Page

158 *According to a Columbia University study:* National Center for Disaster Preparedness, Mailman School of Public Health, *How Americans Feel About Terrorism and Security: Two Years After 9/11* (New York: Columbia University, 2003): 10, www.ncdp.mailman.columbia.edu/How_Americans_Feel_About_Terrorism.pdf.

158 *As Senator Warren Rudman has warned:* James Dao, "Threats And Responses: Domestic Security; U.S. Vulnerability to Terror Is Still High, Panel Concludes," *The New York Times*, October 26, 2002.

162 *He publicly released a lessons-learned report:* For more information about the New York City August 2003 blackout

see New York City Emergency Response Task Force, *Enhancing New York City's Emergency Preparedness: A Report to Mayor Michael R. Bloomberg* (New York: October 28, 2003).

163 *Americans would do well to reflect: Oxford Reference Online*, "Japanese American Internment," www.oxfordreference.-com/views/ENTRY.html?subview=Main&entry=t63.e1007.

163 *Rather, this decision was supported by President Franklin D. Roosevelt:* On December 18, 1944, Justice Black delivered the opinion of the U.S. Supreme Court in *Toyosaburo Korematsu v. United States*, 323 U.S. 214, 65 S.Ct. 193, www.weslaw.com.

163 *In other words, this gross injustice:* James Woolsey made this point during his keynote address at "The Long War of the 21st Century: How We Must Fight It" (The Conference Board, 2003 Corporate Security and Crisis Management Conference: Emerging Issues, Waldorf Astoria Hotel, NY, November 4, 2003).

Index